Acknowledgements

Dedicated to the Memory of Close Friend and Fellow Anarchist, Cady Hine,

The Most Brilliant α I have Ever Met, for Inspiration and ϵ-correction: $B_{\mathbb{F}_{SCHO}}([\alpha\nu])$, $\bigotimes$, $\mathcal{H}^{\infty}_{\alpha i \rho}$, $J^{\mathbb{Z}}$,

My Brightest and Most Loyal ϵ for β-testing,

Of course, Leon Simon, for his Incredible Vitality and Enthusiasm,

And All Mathematical Underdogs, Future, Past, and Present.

$$\begin{bmatrix} 9 & 5 & 4 & 3 & 1 & 0 & 0 \\ 0 & 4 & 3 & 5 & 8 & 0 & 1 \\ 0 & 3 & 0 & 0 & 3 & 0 & 9 \\ 0 & 2 & 0 & 1 & 0 & 5 & 6 \\ 0 & 2 & 1 & 0 & 0 & 4 & 0 \end{bmatrix}$$

Contents

Preface

I pass, like night, from land to land;
I have strange power of speech;
That moment that his face I see,
I know the man that must hear me:
To him, my tale, I teach.

-Rime of the Ancient Mariner

0.1 Three Tales

I've never been that great of a writer. A writer can create originality: original mathematics. I am more of a storyteller: when I love a proof, I can rederive it with my own original spin. My goal today is to be a storyteller and to tell three tales.

The first tale takes place on a warm October night. The crowd was shuffling into Cubberly Auditorium, eagerly awaiting Brian Conrad's public lecture, "Rubik, Escher, Bank." For a mathematics lecture, this was a pretty diverse crowd, filled with students and professors alike. In walked a freshman, who took a seat next to me. Recognizing me from his Honors Multivariable Course, he inquired, with a grin,

I wonder how many of us are here from Math 51H?

This question was rhetorical. Even in the midst of midterm week, it was evident that the front row was teeming with frosh. Being the oddball that I am, I gave a cryptic reply:

Many. And not just your year.

Indeed, the young faces in the audience spanned at least a decade of upperclassmen, doctorates, and graduates who all had one thing in common: they had all been through the introductory Math Honors series. In this freshman's eyes, the presence of his peers was a testament to the strength and passion of his class. But in my eyes, **the presence of so many H-series veterans was a testament to Leon Simon's strength as a teacher.** I have taught for 7 years, across 3 states and 2 continents, and I have never met a more interesting and enthusiastic teacher than Leon Simon. As a teacher, I know there is no greater honor than to see one's students be inspired to continue the cause. And in the small window I have been at Stanford, I have seen some of the brightest and most mathematically talented emerge from his doors.

But not every student can survive the rigors of the H-series. Not every Dante gets out of the Inferno. And that leads me to my second story.

The second tale is from my youth. Truth be told, I used to be a pretty bad kid. And then a professor emeritus took me under his wing. Four years of teaching and he didn't ask for a dime. He was the greatest man I have ever met and more of a father than I have ever known. All I ever really wanted was to be just like him: a mathematician. I guess that's a pretty lousy reason. And I've realized that I was a pretty naive kid: math is more than just Calc BC and the pages of Stewart and Strang.

During my first week of Stanford, my undergraduate advisor asked me one key question to test my math abilities:

What is the definition of continuity?

I gave him the typical American high school answer, epsilon and delta free. And he scolded me. He was right to do so. But before I walked out, he did impart some words of wisdom:

Baptisma Pyros. Be prepared for a Baptism by Fire.

And it was. My undergraduate career was a Baptism by Fire.

The final tale is a retelling of one of the greatest stories ever told, Math 51H. It will mimic the released notes exactly, except I have added additional commentary to build intuition, outlined proofs and techniques, and gave numerous examples and analogies.

But before I begin the tale, I should explain a few things.

0.2   Insight on your Present, Past, and Future

More than a quarter of the 51H class will have solid proof backgrounds and more than half will have made it to AIME. At least ten kids are going to be from highly competitive math camps like PROMYS and SUMaC. Others will hail from countries like Romania, Bulgaria, and Singapore, which have relentless math programs. These students will know all sorts of obscure theorems like which odd primes can be written as a sum of two squares. They will swim through the class like water and be the ones raising their hands. But **every students in the class** will be from the top of his or her school and will want to be a mathematician.

Here is a glimpse of your future:

You will study hard for the first test. It's probably your first Stanford midterm. It's also the first exam you have ever taken at night. You feel confident afterwards because you have aced every math exam ever thrown at you. Then you get the results:

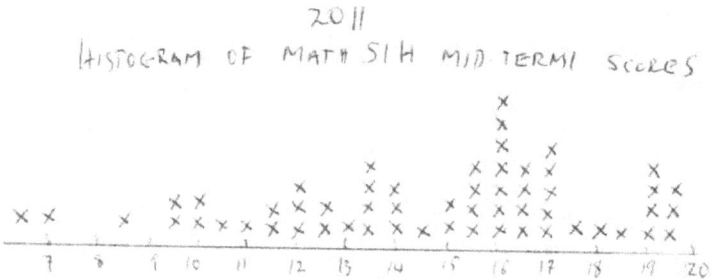

WHAT THE F**K?! You are shocked at the histogram spread. You are even more shocked that you are lower than the 50th percentile. Pretty pissed at yourself, you study *even harder* for the second midterm.

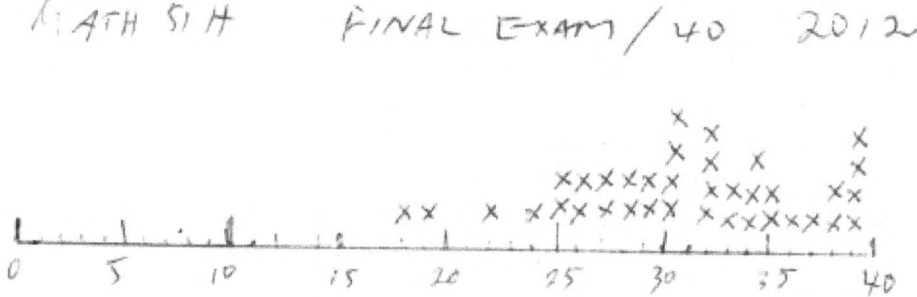

The students who completely bombed the first midterm dropped out. Even though you studied more, you did *even worse*.

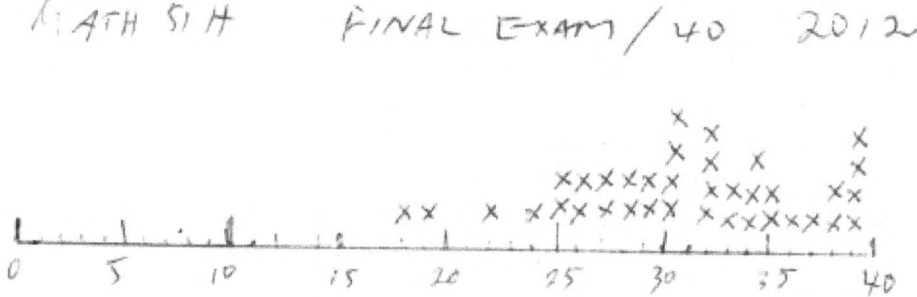

On the final, you score less than a 30 and walk out of the class with a B. You don't admit it to anyone, and if anyone asks, you got an A. You spend your winter break studying for 52H, and when you return, you realize all the B and lower students dropped out. Fifty students become twenty. The cycle begins again, except this time you are dead last.

Is it Leon Simon's fault that the class is going way too fast? No. But how can you master proof techniques like induction if you are flung head-first into upper-level applications? The truth is, Leon Simon cannot cater to everyone. There are so many different math backgrounds spanning the *entire globe*, and ultimately it is **your choice to be in the course**. And it is your misfortune that you went to a normal American high school. Because I have a confession. On behalf of all high school math teachers, I want you to know:

I am sorry. Math is more than just SAT and calculation. It is an art whose beauty and creativity have been omitted from the standard curricula. If you come from the typical American high school, you are at an absurd disadvantage.

I can write a whole book about how bad the mathematics in America is, from elementary rote memorization to mindless exercises.[1] Paul Lockhart already beat me to it in his "A Mathematicians's Lament" (I am sure Professor Devlin would be willing to lend you a copy). There is nothing I can do to reform your education. But at the very least, I am going to give you a fighting chance at the H-series.

0.3   Who Should Read this Book?

Do not read this book if any of the following are true:

- **If you want to be an engineer**. The H-series will actually hurt you. **You must practice calculation**. For any linear dynamical system, Fourier transform, infinite summation, or overdetermined system of equations, you will be asked to forsake any theoretic considerations (like convergence and existence) and just mindlessly calculate. You would be surprised how many theoretical math majors cannot calculate curls or solve simple ODE's. Heck, I remember Leon Simon once asking the class to use integration by parts, and everyone just stared in awkward silence. If you want to understand what's *underneath the box*, then **take Math 115 and Math 109**: they count for the Engineering Mathematics requirement as well as an easy minor.

- **If you are already well-versed in math proofs**. You have to get used to reading concise[2] proofs and figuring out the **why** on your own. Especially when you hit the yellow and blue graduate texts (these books are beasts)!

- **If you are used to taking the highest available math course and you just want a good grade**. First, you are not going to get an A with that attitude. And second, from someone who works in the real world, **no one cares** about the specific classes you take.

- **If you want to make the climb without a rope**. Like Bruce Wayne in "Dark Knight Rises," fear can be a powerful motivating factor. If this is the case, I wish you the best of luck.

Now, if you satisfy the following **necessary** conditions

- **You feel, deep in your core, that you want to be a mathematician.**

- **You want to think about this stuff way after graduation.**

- **You think proofs are cool, and that mathematics is a beautiful art.**

- **You do not have the same math background as everyone else and you always feel like you're at the bottom of the class.**

Then, at this point, I recommend heading to Coursera and watching Keith Devlin's

Introduction to Mathematical Thinking

---

[1] Be prepared to learn the difference between exercises and problems.
[2] To quote Professor Simon, I really need to "trim the fat" from this book.

as soon as possible! In fact, if you are really eager to become a mathematician, chances are you already watched this series before coming to Stanford. Nevertheless, it will get your mind rolling in the right direction.

Then, instead of the H-series, you should take the following courses:

- **Math 115, Real Analysis:** An excellent and easy introduction to the inner workings of Calculus.

- **Math 109, Group Theory:** Here, you will get tremendous practice in deriving theorems from completely abstract properties.

- **Phil 151, Introduction to Logic:** A first course on purely deductive reasoning in a Logic System, an axiomatic view of mathematical reasoning.

- **CS 103, Discrete Mathematics:** Induction, Pigeon-hole, and lotsa fun stuff.

- **Math 110, Number Theory:** Cool properties of primes, modular arithmetic, and RSA.

The first 2 weeks of fundamentals taught in Math 51H will be spread out across 10 weeks in each of the aforementioned courses.

The final necessary condition to read this book is

You are an incredibly stubborn bastard.

Chances are, you are just as stubborn as I was. And you refuse to give up. You will spend your days studying your butt off, stuck in a library, and your grades won't reflect the effort you put in. **How can 10 weeks of work possibly be reflected in 6 hours of testing?** Then this is absolutely the book for you. And if this book helps you in any way, then the time I spent writing it (not to mention the risk of getting sued) was worth it.

0.4   What is a Mathematician?

In the whole discussion above, I never really defined the term mathematician. How can you decide to be a mathematician if you don't even know what one is? In general, there are a few questions for which math people have automated responses. When discussing irrationality, they cite the drowning of Hippasus by the Pythagorean cult. On the topic of constructing the reals from the integers, they always cite Kronecker. To define a mathematician, they quote Hardy:

A mathematician, like a painter or poet, is a maker of patterns. If his patterns are more permanent than theirs, it is because they are made with ideas.

-G.H. Hardy

This is a great answer. However, this is not mine. My answer took me a very long time to find. However, I am not going to tell you, but simply point you where to look. Study the counterexample to an infinitary extension of component convergence implies point convergence of $\mathbb{R}$ to $\mathbb{R}^\infty$. In the words of the great mathematician Leon Simon, who I will often quote in this book,

Mull it over.

0.5   Final Words of Advice

What you **should not** be doing:

- **Skipping the proofs and understanding only calculation.** Every problem on your exam (except 1 definition and 1 calculation), will be a proof

- **Highlighting every word in a proof to memorize an argument exactly.** If you are memorizing rather than understanding, I guarantee you will bomb the course.

- **Not attending lecture**. This is where you gain intuition on proofs and problem solving techniques. Not to mention, Professor Simon is highly entertaining.

- **Wikipedia-ing and googling solutions**. This is like asking someone to lift your own weights! Don't do it. The problem set is the only time you get to practice problem solving, and honestly, you should be spending the *whole week* mulling over the p-set. In the worst case scenario, ask Professor Simon or the TA for a hint. By the way, anyone who tells you that he finished the p-set in less than an hour is either a liar or has seen the material before.

- **Hating yourself if you are completely lost in lecture.** You are seeing it for the first time: of course you will be lost on the spot! To quote Leon Simon,

The $\epsilon - N$ definition took 100 years to develop,
yet you are expected to know it in less than 20 minutes!

If it makes you feel any better, even the great Hilbert, in his youth,

...was not particularly quick at comprehending new ideas. He seemed never really able to understand anything until he had worked it through his own mind.

-Constance Reid, *Hilbert*

What you should be doing:

- **Rewriting and rederiving proofs without looking**. Do this even if you think you know a proof because,

The human capacity for self delusion is limitless

-Leon Simon

In fact, for each of the key theorems, I've written a **proof summary**. Use this first to get an overall picture of the proof and then try to fill in the blanks.

- **Going to office hours.** Ask questions, even if it makes you look stupid. To quote one of the math department's top graduate students,

In undergrad, I pestered a professor with questions after every lecture, and initially I felt he wanted to escape me; later he said he was very happy to receive questions because it showed someone was listening to and understanding his lecture. He ended up teaching me tons of hard maths beyond his course, and wrote me an excellent recommendation letter, both of which really helped me get into grad school. So you might get something amazing out of talking to your professors; you wouldn't know if you didn't try.

-Amy Pang

- **Improving your proofs even if you already understand them**: it is like improving a Pina Colada by adding amaretto. This is especially true when homework solutions are released. Even if you get the proofs correct, read over the solutions!

- **Being in a state of Sitz-Fleisch**. This is the first thing Leon Simon will tell you (and one of my favorite expressions). You have to be in a state of "sitting-flesh," constantly thinking about a problem and expanding your mind.

Now, without further ado, my re-telling of Math 51H. I hope you find it more useful than the Half Blood Prince's annotated copy of advanced potion making.

Lecture 1

Distance, Defined

The journey of a thousand miles begins with some sort of metric.

- ⇐cius

Goals: The first three weeks of 51H are dedicated to Linear Algebra: this is because the objects we will be working with are vectors. Today, we define what a vector is, as well as the distance between two vectors. We also introduce sets and the method for proving universal statements.

1.1   Know thy Enemy

Before we begin our ten-week journey, we first need to know what we're studying. Particularly, we need to ask ourselves,

What is Multivariable Calculus?

This is a two-part question: first, what is *Calculus*?

In high school, you learned that

*Calculus is the study of **change**.*

However, I feel a more apt description is that

*Calculus is the study of **closeness**.*

This is because in Calculus, you study sequences as the term number gets "closer to infinity"

$$\lim_{n \to \infty} a_n$$

difference quotients over interval lengths that get closer to 0

$$\lim_{h \to 0} \frac{f(x+h) - f(x)}{h}$$

and area approximations that get closer to the true value:

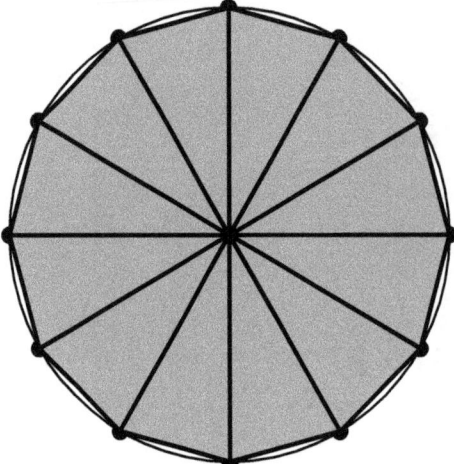

But to study the nature of *closeness*, we need a notion of **distance** between **objects**. And the type of objects we will be dealing with is what makes this *Multivariable* Calculus. Namely, we will be working with **vectors**.

1.2   Vectors

In your high school career, you worked with single numbers x, pairs (x, y), and triplets (x, y, z). However, we do not need to limit ourselves to just 1, 2, or 3 components. In this course, we will generalize to n components:

$$(x_1, x_2, \ldots, x_n).$$

Most of the time, we will prop these n-tuples as columns:

Definition. *An n-dimensional (column[1]) **vector** is an n-tuple*

$$\vec{v} = \begin{bmatrix} v_1 \\ v_2 \\ \vdots \\ v_n \end{bmatrix}$$

where $v_1, v_2 \ldots, v_n$ are real numbers.

By convention, variables with an overhead arrow will denote vectors. Moreover, given a vector $\vec{x}$, we will denote its i-th component by x_i. Thus,

$$x_1, x_3, x_7$$

will represent the 1st, 3rd, and 7th components of $\vec{x}$, respectively.

---

[1]We will always assume a vector is a column vector unless stated otherwise.

Algebraically, we can add and subtract vectors of the same size

$$\vec{v} + \vec{w} = \begin{bmatrix} v_1 \\ v_2 \\ \vdots \\ v_n \end{bmatrix} + \begin{bmatrix} w_1 \\ w_2 \\ \vdots \\ w_n \end{bmatrix} = \begin{bmatrix} v_1 + w_1 \\ v_2 + w_2 \\ \vdots \\ v_n + w_n \end{bmatrix}$$

and scale vectors by a constant

$$c\vec{v} = c \begin{bmatrix} v_1 \\ v_2 \\ \vdots \\ v_n \end{bmatrix} = \begin{bmatrix} cv_1 \\ cv_2 \\ \vdots \\ cv_n \end{bmatrix}.$$

For the cases $n = 1, 2, 3$, you can visualize vectors geometrically as points in space or directed arrows from the origin:

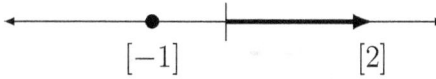

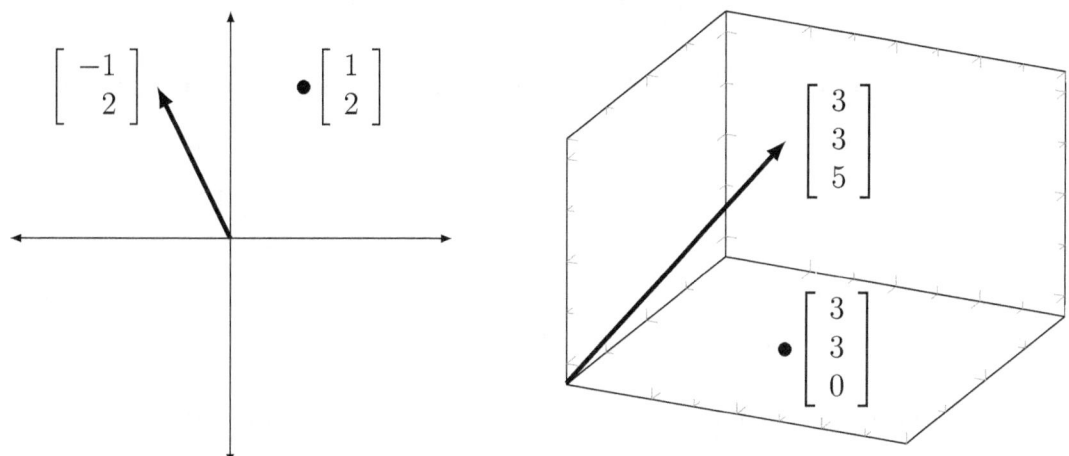

But how do you visualize the case $n = 4$ or higher?

I have heard some physics mumbo jumbo of visualizing four or higher spatial dimensions. I really don't get it. If you can see in 4D, then you are too gifted to be taking undergraduate math. Regardless, there are *many* ways you can interpret n-dimensional vectors.

For example,

$$\vec{x} = \begin{bmatrix} 2 \\ 3 \\ 4 \\ 5 \end{bmatrix}$$

can represent a particle at $(2, 3, 4)$ at time 5:

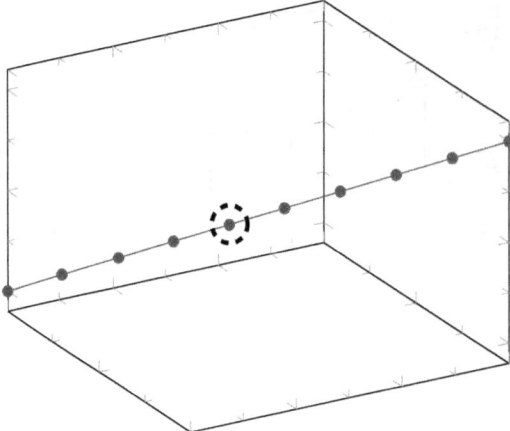

But we can also interpret the same vector as a colored point on the number line:

$$\vec{x} = \begin{bmatrix} 2 \\ 3 \\ 4 \\ 5 \end{bmatrix} \begin{matrix} \longleftarrow \text{Position} \\ \longleftarrow \text{Red Quantity} \\ \longleftarrow \text{Blue Quantity} \\ \longleftarrow \text{Green Quantity} \end{matrix}$$

Therefore, treat vectors as purely *algebraic objects* and leave the interpretation to the *context*.

1.3   Sets

Often times, we would like to talk about some collection of vectors. In order to do this, we need to discuss one of the most fundamental[1] objects of mathematics: *sets*.

A *set* is simply a collection of *distinct* objects. The most common sets that you will see in your undergraduate career are:

Symbol	Definition
$\mathbb{Z}$	The set of all integers.
$\mathbb{Q}$	The set of all rationals.
$\mathbb{R}$	The set of all real numbers.
$\mathbb{C}$	The set of all complex numbers.
$\mathbb{R}^n$	The set of all n-dimensional vectors.

We can also explicitly build sets using curly brackets:

$$A = \left\{ \dots \right\}$$

---

[1]In **Math 161: Set Theory**, you will learn that we can encode mathematics using sets.

For example,

$$A = \{\text{Red}, \text{Green}, \text{Blue}\}$$
$$B = \{\vec{x}_1, \vec{x}_2, \vec{x}_3, \vec{x}_4, \vec{x}_5\}$$

tells us that A is the set containing the colors Red, Green, and Blue and that B is the set containing the vectors $\vec{x}_1, \vec{x}_2, \vec{x}_3, \vec{x}_4, \vec{x}_5$.

Of course, we can have more complicated sets. Particularly, we can form sets by picking out elements from a bigger set. In this case, we will use the notation

$$A = \Big\{ x \in B \ \Big| \ P(x) \Big\}$$

which is read as

A is the set of all elements x in B such that x satisfies property P

For example,

$$S = \Big\{ x \in \mathbb{Z} \ \Big| \ x = y^2 \text{ for some } y \in \mathbb{Z} \Big\}$$

is the set of all perfect squares whereas

$$T = \Big\{ \vec{x} \in \mathbb{R}^5 \ \Big| \ x_1 = 0 \Big\}$$

is the set of all 5-dimensional vectors with first component 0.

1.4   Distance

Now that we have the objects of Multivariable Calculus, we would like to measure the distance between them.

But what is the distance between two vectors?

In high school, you derived the distance formulas for 2D and 3D by repeated application of the Pythagorean Theorem:

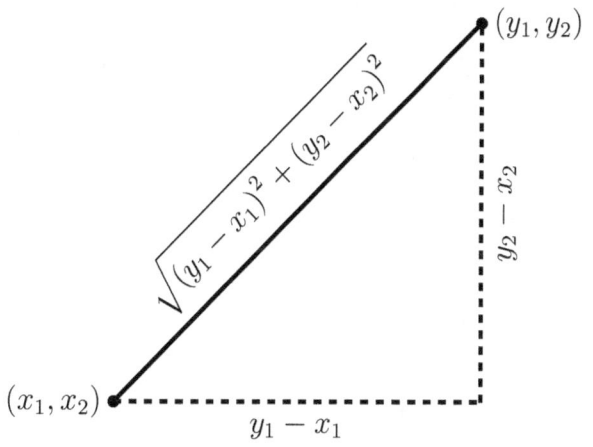

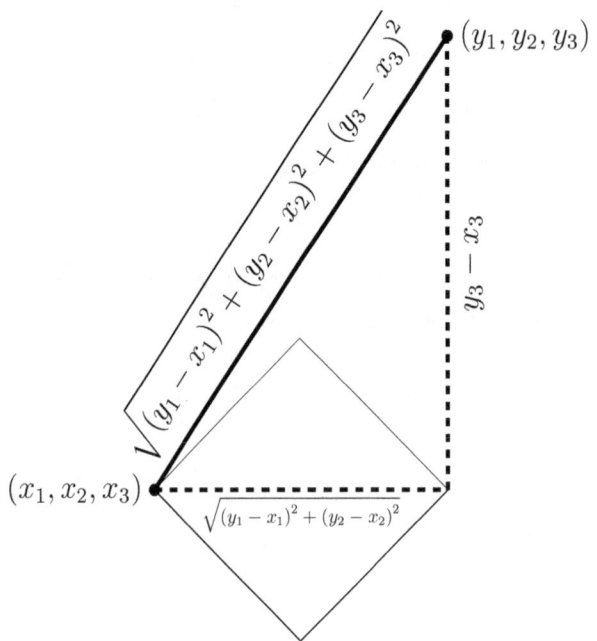

From these two cases, you could hypothesize that

Distance can be calculated by plotting the two vectors spatially and
repeatedly applying the Pythagorean Theorem.

But this is incorrect! A vector is an *algebraic construct,* while the Pythagorean Theorem is a geometric statement about lines in the plane. We can't even visualize higher dimensional vectors!

Instead, we must *define* the distance formula for vectors. But where do we begin?

We can't just spout out any formula. Our formula needs to have *meaning.* Namely, it must capture the real world intuition of distance.

Generally,

> Math Mantra: Study a physical phenomena and decide on its key properties.
>   Then, incorporate these properties in an abstract definition.

We *decide* that distance should be a *function* that inputs two vectors and outputs a real number. Moreover, we *decide* that a distance function should have the following properties:

Definition. *We call a function* $d : \mathbb{R}^n \times \mathbb{R}^n \to \mathbb{R}$ *a* ***distance function*** *on* $\mathbb{R}^n$ *if the following properties hold:*

1. ***(Non-negativity)*** *Distance is always non-negative: for all* $\vec{x}, \vec{y} \in \mathbb{R}^n$,

$$d(\vec{x}, \vec{y}) \geq 0.$$

2. ***(Symmetry)*** *The distance from* $\vec{x}$ *to* $\vec{y}$ *should be the same as the distance from* $\vec{y}$ *to* $\vec{x}$:

$$d(\vec{x}, \vec{y}) = d(\vec{y}, \vec{x})$$

for all $\vec{x}, \vec{y} \in \mathbb{R}^n$.

3. ***(Zero)*** *If the distance between two objects is 0, then they should be the same object: for all* $\vec{x}, \vec{y} \in \mathbb{R}^n$, *if*

$$d(\vec{x}, \vec{y}) = 0$$

then

$$\vec{x} = \vec{y}.$$

4. ***(Triangle Inequality)*** *The direct path between two vectors should be shorter than or the same length as any detour:*

$$d(\vec{x}, \vec{z}) \leq d(\vec{x}, \vec{y}) + d(\vec{y}, \vec{z})$$

for all $\vec{x}, \vec{y}, \vec{z} \in \mathbb{R}^n$.

This is an ideal definition for a distance function; however, we need to be careful:

Math Mantra: A definition does not guarantee existence!

How do we know there is actually a function d that satisfies all four properties? For example, I can make the definition,

Let n *be a number that is both even and odd.*

But no such n exists! Likewise, I can say

Let x *be a good M. Night Shyamalan movie.*

and as we all know, there is no such x.

To find a suitable candidate for a distance function d, we look to the 2D and 3D distance formulas for inspiration:

$$\sqrt{(x_1 - y_1)^2 + (x_2 - y_2)^2}$$

$$\sqrt{(x_1 - y_1)^2 + (x_2 - y_2)^2 + (x_3 - y_3)^2}.$$

Naturally, since we are working with n-dimensional vectors, we guess that

$$d(\vec{x}, \vec{y}) = \sqrt{(x_1 - y_1)^2 + (x_2 - y_2)^2 + \ldots + (x_n - y_n)^2}$$

is a distance function. But in order to verify this, you'll need a basic proof technique:

1.5   Proof Technique: Proving Universal Statements

Suppose you want to prove a fact about every element in a set.

For example, you may want to prove

Every natural number can be written as a sum of four squares.

Or, as a less abstract example,

Every Californian bar is forbidden to sell Everclear-190.

The bone-headed thing to do (if it is even possible) would be to go through each particular case. In our second example, this means *bothering* every bar owner in California.

The more logical thing to do is to take an **arbitrary** element in set S and call it x:

$$x \in S.$$

Then, **using only the fact that $x \in S$**, show that x has the property you are trying to prove. Since we can reapply this argument for any particular element in S, we can conclude **all elements in S must have this property.**

So in our second example, consider an arbitrary bar in California and call it x. Then we use the following chain of reasoning:

- By definition of being a Californian bar, x has to have a Californian liquor license.

- A Californian liquor license only permits the sale of alcoholic beverages with alcoholic volume of at most 75.5%.

- Everclear-190 has volume 95%.

- Thus, x cannot sell Everclear-190.

But we only used the fact that x is a Californian bar and not its particular identity. Thus, we can conclude that *any* bar in California we choose is forbidden to sell Everclear-190. In other words, *every* Californian bar is forbidden to sell Everclear-190.

Here are a few mathematical examples:

Example. *Every odd number can be written as a difference of two squares.*

Proof: Let x be an arbitrary odd number. By definition,

$$x = 2n + 1$$

for some integer $n \in \mathbb{Z}$. Now consider the consecutive perfect squares, $(n+1)^2$ and n^2. Then,

$$(n+1)^2 - n^2 = 2n + 1 = x.$$

Thus, x can be written as a difference of two squares. Since x was an arbitrary odd number and we only used the fact that x was odd, we conclude that all odd numbers can be written as a difference of squares. $\square$

Notice, in this proof, that the choice of n is a *function* of the arbitrary choice x. If we picked $x = 5$, then we would know $n = 2$. Or, if we picked $x = 301$, then $n = 150$. The point is, we can construct n for any choice x.

Example. *The sum of two rational numbers is always rational.*

Proof: Let a, b be two rational numbers. By definition of being rational, there exist integers p_1, p_2, q_1, q_2 such that

$$a = \frac{p_1}{q_1}$$

$$b = \frac{p_2}{q_2}$$

Then,

$$a + b = \frac{p_1}{q_1} + \frac{p_2}{q_2}$$

$$= \frac{p_1 q_2 + p_2 q_1}{q_1 q_2}$$

Since $p_1 q_2 + p_2 q_1$ and $q_1 q_2$ are still integers, $a + b$ is rational. Moreover, the choice of a, b was arbitrary; thus, the sum of any two rational numbers is rational. $\square$

In the preceding proof, you need to avoid a noob mistake:

Math Mantra: DON'T MAKE A DUMMY MISTAKE WITH DUMMY VARIABLES!

Suppose you wrote:

Since a, b are rational,

$$a = \frac{p}{q}$$

$$b = \frac{p}{q}$$

for some integers p, q.

This is **completely wrong**: this asserts that $a = b$.

True, there is no problem in changing the dummy variable. For example,

$$\sum_{i=1}^{N} i$$

is the same as

$$\sum_{j=1}^{N} j$$

However, if you had used the same dummy variables in the preceding theorem, you would have completely changed the *intent* of the expression. Generally,

> **Math Mantra:**   The purpose of notation is to precisely capture our mathematical reasoning.   It is the mathematics that create the notation, NOT the other way around.

For our final example, recall that a function f is *even* if

$$f(x) = f(-x)$$

for every x. Likewise, a function f is *odd* if

$$f(x) = -f(-x)$$

for every x. While every integer is either even or odd, some functions are neither even nor odd. But it turns out that every function has a neat decomposition:

Example. *Any function can be written as the sum of an even function and an odd function.*

Proof: Consider an arbitrary function f. Using f, we can create functions

$$g(x) = \frac{f(x) + f(-x)}{2}$$

$$h(x) = \frac{f(x) - f(-x)}{2}$$

Notice that g is even since

$$g(-x) = \frac{f(-x) + f(-(-x))}{2} = \frac{f(-x) + f(x)}{2} = g(x).$$

Moreover, h is odd:

$$h(-x) = \frac{f(-x) - f(-(-x))}{2} = \frac{f(-x) - f(x)}{2} = -\frac{f(x) - f(-x)}{2} = -h(x).$$

Summing g and h, we get

$$g(x) + h(x) = \frac{f(x) + f(-x)}{2} + \frac{f(x) - f(-x)}{2} = \frac{2f(x)}{2} = f(x)$$

Hence, f can be written as the sum of an even function and an odd function. Since f was arbitrary, any function can be written as the sum of an even function and an odd function. □

Here, you should be asking,

Where in the world did g and h come from?

The truth is, we had to *find* these explicit functions. And despite appearances, these functions didn't pop out of thin air: we needed to do lots of scratch work.[1]

Math Mantra: Most mathematics comes from being locked in a room and wasting
tons of paper.

And when you find a beautiful idea, you keep this single sheet and burn all evidence of failure. This is what makes mathematics an art.

1.6   How Proofs Should Not be Done

As an application reader for a prestigious mathematics program, I see a lot of top students apply. All of them with 4.0 GPA, 10+ AP's, perfect 800 SAT Math scores, great letters of recommendation, and long-winded essays. And to be honest, more than half of them submit atrocious problem sets.

Consider the problem:

Prove that $n^2 + n + 1$ is never a multiple of 5.

More than half of the students submit an answer like this:

---

[1]Although guessing an explicit formula is a tough process, experience will guide your path. After some exposure to Complex Analysis and Euler's formula, thinking up g and h will become second nature.

> I tested 1~15 for 'n' in the equation, n²+n+1, and this
> is what I got: With these results, I found that the
>   1│ 3   11│133   units places went in a pattern,
>   2│ 7   12│157   repeating after 1~10. (I've
>   3│13   13│183   highlighted it) Since all
>   4│21   14│210   multiples of 5 have a units
>   5│31   15│241   digit of either 0 or 5, I
>   6│43   - - - - - - - -   concluded that there is no
>   7│57   value for 'n' where n²+n+1 is a multiple
>   8│73   of 5.
>   9│91   ┌─────────────────────┐
>   10│111   │ ∴ Answer: none   │
>   └─────────────────────┘

This is not a proof! It is not enough to plug in a few numbers!

For example, suppose you wanted to "prove"

$$p(x) = x^2 + x + 41$$

spits out a prime number for any integer x. The completely boned-headed thing to do would be to plug in the integers 0, 1, 2, 3, and then conclude it is true. In fact, if you keep on plugging in integers from 1 to 39, you will keep getting a prime. But just plug in 40, and you will see

$$p(40) = 1641$$

isn't prime[1]!

Let's be honest: you can tell this is not a valid proof, but many of you would have submitted an answer like this. Does this mean you are stupid? **No.** This is what **we taught you.** But now I am teaching you something different:

```
       Math Mantra:  Proof does not mean ``support the argument!''
```

A proof is an undeniable, irrefutable, absolutism. It is the final word on all arguments, the Judge Dredd of the sciences. It is my hope that this book not only helps gets you through the H-series, but also helps correct High School misconceptions.

Before we return to the main theorem, we need one more digression on a very important fact.

1.7   Squaring and Rooting

In this course, we will be working with inequalities. **A lot.** Unfortunately, there are few high school exercises that work with inequalities abstractly. In particular, you may have missed two key facts involving **non-negative numbers:**

---

[1]If you want more curios like this, Google search the "Law Of Small Numbers" by Richard Guy. FYI, there does not exist a non-constant polynomial with integer coefficients that always outputs a prime number.

Square roots preserve inequalities.

and

Squaring both sides preserves inequalities

Formally, this means that $\sqrt{x}$ and x^2 are increasing functions over the non-negative numbers: if

$$0 \le a \le b$$

then

$$\sqrt{a} \le \sqrt{b}$$

and

$$a^2 \le b^2$$

For example,

$$3 \le 9$$

thus

$$\sqrt{3} \le \sqrt{9}$$
$$3^2 \le 9^2$$

Note that we *need* the non-negativity requirement since, for negative numbers,

1. $\sqrt{x}$ is undefined

2. Squaring would not preserve inequalities. For example,

$$-5 \le 3$$

but

$$25 \ge 9.$$

You can check that x^2 and $\sqrt{x}$ are increasing functions using Calculus, but if you don't believe me, just stare at their graphs:

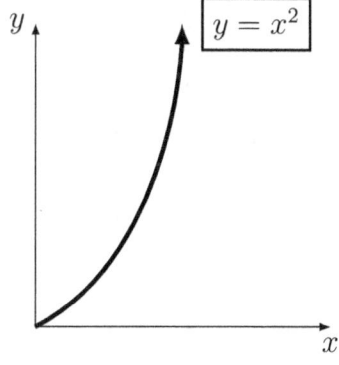 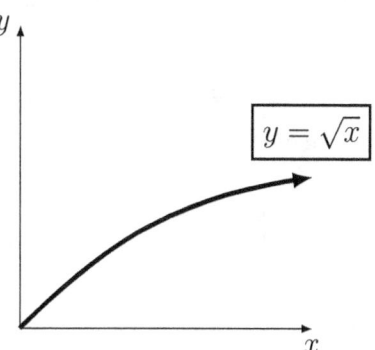

Now, back to our regularly scheduled program.

1.8   Verifying that d is Actually a Distance Function

Theorem. *The function $d : \mathbb{R}^n \times \mathbb{R}^n \to \mathbb{R}$ defined by*

$$d(\vec{x}, \vec{y}) = \sqrt{(x_1 - y_1)^2 + (x_2 - y_2)^2 + \ldots + (x_n - y_n)^2}$$

satisfies the properties of a distance function.

Proof: We simply check the definition:

- *Non-negativity:*

  Let $\vec{x}$ and $\vec{y}$ be two arbitrary vectors in $\mathbb{R}^n$. Since a sum of squares is always non-negative,

  $$(x_1 - y_1)^2 + (x_2 - y_2)^2 + \ldots + (x_n - y_n)^2 \geq 0.$$

  Moreover, square roots preserve inequalities; thus, we can root both sides to get

  $$\sqrt{(x_1 - y_1)^2 + (x_2 - y_2)^2 + \ldots + (x_n - y_n)^2} \;\geq\; 0$$

  which is the same as

  $$d(\vec{x}, \vec{y}) \geq 0.$$

- *Symmetry:*

  Simply use the fact that $x^2 = (-x)^2$: for arbitrary $\vec{x}, \vec{y}$,

  $$\begin{aligned}
  d(\vec{x}, \vec{y}) &= \sqrt{(x_1 - y_1)^2 + (x_2 - y_2)^2 + \ldots + (x_n - y_n)^2} \\
  &= \sqrt{(y_1 - x_1)^2 + (y_2 - x_2)^2 + \ldots + (y_n - x_n)^2} \;= d(\vec{y}, \vec{x})
  \end{aligned}$$

- *Zero:*

  Assume

  $$\sqrt{(x_1 - y_1)^2 + (x_2 - y_2)^2 + \ldots + (x_n - y_n)^2} = 0$$

  Squaring, we get

  $$(x_1 - y_1)^2 + (x_2 - y_2)^2 + \ldots + (x_n - y_n)^2 = 0$$

  But the sum of non-negative terms is 0, when, and only when each term is 0:

  $$\begin{aligned}
  (x_1 - y_1)^2 &= 0 \\
  (x_2 - y_2)^2 &= 0 \\
  &\vdots \\
  (x_n - y_n)^2 &= 0
  \end{aligned}$$

  which is only true when

  $$\begin{aligned}
  x_1 &= y_1 \\
  x_2 &= y_2 \\
  &\vdots \\
  x_n &= y_n
  \end{aligned}$$

  Thus, $\vec{x} = \vec{y}$.

- *Triangle Inequality:*

  See next lecture. □

Note, we have yet to prove that $d(\vec{x}, \vec{y})$ is a distance function (although we will finish the proof in the next lecture, as a result of Cauchy-Schwarz inequality).

Because a binary function $d(\vec{x}, \vec{y})$ is cumbersome to write, we define (for this entire course),

---

Definition. *The **Euclidean distance**[1] between $\vec{x}, \vec{y} \in \mathbb{R}^n$ is*

$$\|\vec{x} - \vec{y}\| = \sqrt{(x_1 - y_1)^2 + (x_2 - y_2)^2 + \ldots + (x_n - y_n)^2} \quad .$$

*In particular, the distance between $\vec{x}$ and the **zero vector** $\vec{0}$*

$$\|\vec{x}\| = \sqrt{x_1^2 + x_2^2 + \ldots + x_n^2}$$

*is called the **norm** of $\vec{x}$.*

1.9 Why the Distance Function Detour?

I took the liberty of introducing the properties of a distance function before giving an explicit example. The reason is that we can have different functions[2] d that satisfy the properties of a distance function. For example, we could have defined:

$$d(\vec{x}, \vec{y}) = |x_1 - y_1| + |x_2 - y_2| + \ldots + |x_n - y_n|$$

We could have even given d a completely bone-headed definition:

$$d(\vec{x}, \vec{y}) = \begin{cases} 1 & \text{if } \vec{x} \neq \vec{y} \\ 0 & \text{if } \vec{x} = \vec{y} \end{cases}$$

A recurring theme in mathematics is,

Math Mantra: If we prove a theorem using ONLY the properties of an object, then we can apply the theorem to any other object that satisfies the same properties.

The computer scientists call this *using a different instantiation.* So understand the properties of distance, and don't just gloss them over: you'll thank me later.

---

[1]If $\|\vec{x} - \vec{y}\|$ is not a distance function, then this would be a bone-headed name.

[2]In fact, we can take this abstraction one step further: instead of $\mathbb{R}^n$, we can take d over some set M. You will see this in **Math 171** when you study metric spaces.

New Notation

Symbol	Reading	Example	Example Translation
$\in$	In, element of	$x \in A$	x is an element of the set A
$\mathbb{R}$	The set of all real numbers	$\pi \in \mathbb{R}$	π is a real number.
$\mathbb{C}$	The set of all complex numbers	$2 + i \in \mathbb{C}$	$2 + i$ is a complex number.
$\mathbb{Q}$	The set of all rationals	$\frac{1}{2} \in \mathbb{Q}$	$\frac{1}{2}$ is rational.
$\mathbb{Z}$	The set of all integers	$3 \in \mathbb{Z}$	3 is an integer.
$\mathbb{R}^n$	The set of all n-dimensional vectors	$\vec{v} \in \mathbb{R}^2$	$\vec{v}$ is a 2 dimensional vector.
$\left\{ x \in A \mid P(x) \right\}$	The set of all elements in A satisfying property P	$\{ x \in \mathbb{R} \mid x > 0 \}$	The set of all real numbers that are greater than 0.
$\vec{v}$	Vector v	$\vec{v} \in \mathbb{R}^3$	v is a vector with three real components.
$\vec{0}$	The zero vector	$\vec{x} + \vec{0} = \vec{x}$	The sum of vector $\vec{x}$ and the zero vector is $\vec{x}$.
$\|\vec{x}\|$	The norm (or length) of vector x	$\|\vec{x}\| = 1$	The length of vector $\vec{x}$ is 1.

Lecture 2

All About Angles

...there is no doubt that [the Cauchy-Schwarz inequality] is one of the most widely used and most important inequalities in all of mathematics

-J. Michael Steele, *The Cauchy-Schwarz Master Class*

Goals: Today, we define the angle between two vectors. To do this, we introduce the dot product and its properties. Also, to check that our definition of angle makes sense, we derive the Cauchy-Schwarz inequality. The Cauchy-Schwarz inequality is a key inequality which will be used time and time again throughout this course and all of your future analysis courses.

2.1   Angles in $\mathbb{R}^n$

Last lecture, we mentioned that vectors in $\mathbb{R}^2$ and $\mathbb{R}^3$ can be spatially visualized as directed arrows from the origin. Under this visualization, we can calculate the angle between two vectors.

For example, in the case of $\mathbb{R}^2$, suppose we are given the points (x_1, x_2) and (y_1, y_2):

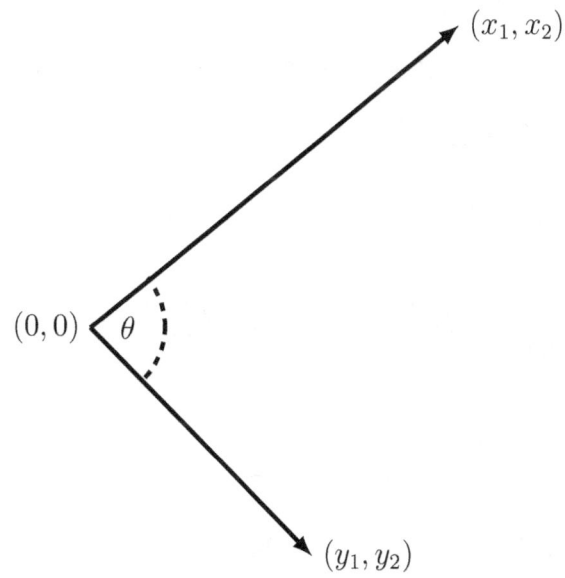

To calculate θ, we first compute the distances:

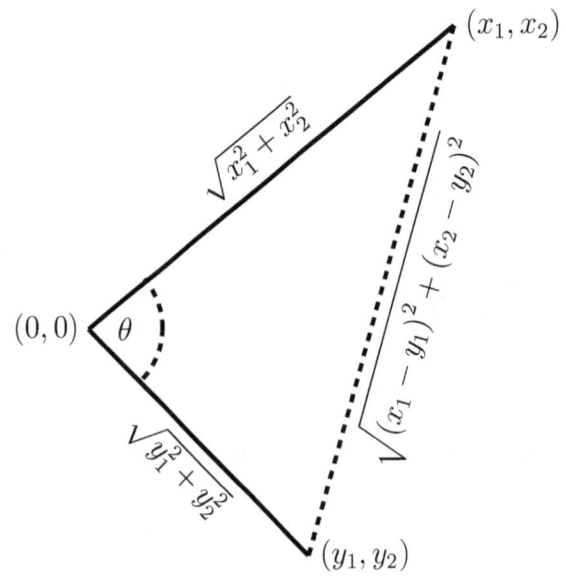

Then, using the Law of Cosines

$$c^2 = a^2 + b^2 - 2ab\cos(\theta)$$

with

$$a = \sqrt{x_1^2 + x_2^2}$$
$$b = \sqrt{y_1^2 + y_2^2}$$
$$c = \sqrt{(x_1 - y_1)^2 + (x_2 - y_2)^2}$$

we have

$$\underbrace{(x_1 - y_1)^2 + (x_2 - y_2)^2}_{c^2} = \underbrace{x_1^2 + x_2^2}_{a^2} + \underbrace{y_1^2 + y_2^2}_{b^2} - 2\underbrace{\sqrt{x_1^2 + x_2^2}}_{a}\underbrace{\sqrt{y_1^2 + y_2^2}}_{b}\cos(\theta).$$

Expanding the left hand side (LHS), we get

$$x_1^2 - 2x_1y_1 + y_1^2 + x_2^2 - 2x_2y_2 + y_2^2 = x_1^2 + x_2^2 + y_1^2 + y_2^2 - 2\sqrt{x_1^2 + x_2^2}\sqrt{y_1^2 + y_2^2}\cos(\theta).$$

Then, cancelling terms, we are left with

$$-2x_1y_1 - 2x_2y_2 = -2\sqrt{x_1^2 + x_2^2}\sqrt{x_2^2 + y_2^2}\cos(\theta).$$

This gives us

$$\cos(\theta) = \frac{x_1y_1 + x_2y_2}{\sqrt{x_1^2 + x_2^2}\sqrt{y_1^2 + y_2^2}}.$$

Thus,

$$\theta = \cos^{-1}\left(\frac{x_1y_1 + x_2y_2}{\sqrt{x_1^2 + x_2^2}\sqrt{y_1^2 + y_2^2}}\right).$$

Likewise, we can follow a similar derivation to show that the angle between (x_1, x_2, x_3) and (y_1, y_2, y_3) is

$$\theta = \cos^{-1}\left(\frac{x_1 y_1 + x_2 y_2 + x_3 y_3}{\sqrt{x_1^2 + x_2^2 + x_3^2}\sqrt{y_1^2 + y_2^2 + y_3^2}}\right).$$

But what about the angle between two vectors in $\mathbb{R}^n$ for $n \geq 4$?

To reiterate, vectors are simply algebraic objects and we cannot visualize $\mathbb{R}^n$ spatially for higher values of n. So just as we defined the distance formula, we will need to *define* the angle between two vectors.

Staring at the 2D and 3D cases,

$$\cos^{-1}\left(\frac{x_1 y_1 + x_2 y_2}{\sqrt{x_1^2 + x_2^2}\sqrt{y_1^2 + y_2^2}}\right)$$

$$\cos^{-1}\left(\frac{x_1 y_1 + x_2 y_2 + x_3 y_3}{\sqrt{x_1^2 + x_2^2 + x_3^2}\sqrt{y_1^2 + y_2^2 + y_3^2}}\right)$$

we are inspired to choose the following candidate for the angle formula in $\mathbb{R}^n$:

$$\cos^{-1}\left(\frac{x_1 y_1 + x_2 y_2 + \ldots + x_n y_n}{\sqrt{x_1^2 + x_2^2 + \ldots + x_n^2}\sqrt{y_1^2 + y_2^2 + \ldots + y_n^2}}\right)$$

If you look closely, you'll recognize that the denominator is the product

$$\|\vec{x}\|\,\|\vec{y}\|.$$

Moreover, the numerator is so important that we give it a name:

Definition. *For $\vec{x}, \vec{y} \in \mathbb{R}^n$, **the dot product** of $\vec{x}, \vec{y}$ is*

$$\vec{x} \cdot \vec{y} = x_1 y_1 + x_2 y_2 + \ldots + x_n y_n$$

Using this notation, we formally **define angles** between vectors in $\mathbb{R}^n$:

Definition. *The angle between two non-zero vectors $\vec{x}, \vec{y} \in \mathbb{R}^n$ is*

$$\theta = \cos^{-1}\left(\frac{\vec{x} \cdot \vec{y}}{\|\vec{x}\|\,\|\vec{y}\|}\right)$$

Again, our definition needs to make sense! Specifically, between any two non-zero vectors, the angle θ should always exist. Suppose

$$\left| \frac{\vec{x} \cdot \vec{y}}{\|\vec{x}\| \, \|\vec{y}\|} \right| > 1$$

Then θ does not exist (the domain of $\cos^{-1}$ is $[-1, 1]$)! Thus, we would like

$$\left| \frac{\vec{x} \cdot \vec{y}}{\|\vec{x}\| \, \|\vec{y}\|} \right| \le 1$$

for $\vec{x}, \vec{y} \ne 0$.

Before we can prove this, we need to learn dot product properties.

2.2   Dot Product Properties

The most basic properties you need to know are:

Theorem. *The dot product satisfies the following properties:*

1. *(**Commutativity**) Order of dot product multiplication does not matter:*

$$\vec{x} \cdot \vec{y} = \vec{y} \cdot \vec{x}$$

   for all $\vec{x}, \vec{y} \in \mathbb{R}^n$.

2. *(**Homogeneity**) You can always pull scalars outside the dot product:*

$$(\lambda \vec{x}) \cdot \vec{y} = \vec{x} \cdot (\lambda \vec{y}) = \lambda(\vec{x} \cdot \vec{y})$$

   for all real λ and $\vec{x}, \vec{y} \in \mathbb{R}^n$.

3. *(**Distributivity**) Dot products distribute over vector addition:*

$$\vec{x} \cdot (\vec{y} + \vec{z}) = \vec{x} \cdot \vec{y} + \vec{x} \cdot \vec{z}$$

   for all $\vec{x}, \vec{y}, \vec{z} \in \mathbb{R}^n$.

4. *(**Relation to Norm**[1]) The dot product of a vector with itself is the same as the square of its norm:*

$$\vec{x} \cdot \vec{x} = \|\vec{x}\|^2$$

   for all $\vec{x} \in \mathbb{R}^n$.

---

[1] In math lingo, we say that the norm is induced by the dot product.

Proof: Expand the definition in each case. For example, to check (1), let $\vec{x}, \vec{y} \in \mathbb{R}^n$. Then, by the commutativity of the reals,

$$
\begin{aligned}
\vec{x} \cdot \vec{y} &= x_1 y_1 + x_2 y_2 + \ldots + x_n y_n \\
&= y_1 x_1 + y_2 x_2 + \ldots + y_n x_n \\
&= \vec{y} \cdot \vec{x} \qquad \qquad \square
\end{aligned}
$$

Notice that the dot product looks a lot like normal multiplication; however, don't think of the dot product *as* multiplication! In particular, associativity doesn't really makes sense here: dot products only take vectors as arguments.

Now that we have these properties, we should

Math Mantra: Try to use the derived properties and not the original definition
to prove a result.

Why shouldn't we go back to the original definition?

1. **It could be tedious.** We already derived the tools, so why not use them right away? Consider the following proof:

   Example. *For any $\vec{x}, \vec{y} \in \mathbb{R}^n$,*

   $$\|\vec{x} + \vec{y}\|^2 = \|\vec{x}\|^2 + \|\vec{y}\|^2 + 2\vec{x} \cdot \vec{y}$$

   Proof: Use the **properties** of dot product to expand the left hand side:

   $$
   \begin{aligned}
   \|\vec{x} + \vec{y}\|^2 &= (\vec{x} + \vec{y}) \cdot (\vec{x} + \vec{y}) & \text{(Convert Norm to Dot Product)} \\
   &= (\vec{x} + \vec{y}) \cdot \vec{x} + (\vec{x} + \vec{y}) \cdot \vec{y} & \text{(Distribute)} \\
   &= \vec{x} \cdot (\vec{x} + \vec{y}) + \vec{y} \cdot (\vec{x} + \vec{y}) & \text{(Commutativity)} \\
   &= \vec{x} \cdot \vec{x} + \vec{x} \cdot \vec{y} + \vec{y} \cdot \vec{x} + \vec{y} \cdot \vec{y} & \text{(Distribute)} \\
   &= \vec{x} \cdot \vec{x} + \vec{x} \cdot \vec{y} + \vec{x} \cdot \vec{y} + \vec{y} \cdot \vec{y} & \text{(Commutativity)} \\
   &= \|\vec{x}\|^2 + \|\vec{y}\|^2 + 2\vec{x} \cdot \vec{y} & \text{(Convert Dot Product to Norm)} \quad \square
   \end{aligned}
   $$

   If we went back to the dot product definition and wrote the norm explicitly as the square root of a sum of squares, it would be a lot messier.

2. **Generalization.** This is the far more important reason which we talked about during the end of last lecture. There are different norms that satisfy properties (1)-(4) in the Theorem. The preceding proof applies to *any* of these norms, whereas a proof involving the explicit definition of the dot product would not.

2.3   Cauchy-Schwarz Inequality

Now that we are equipped with the proper machinery, we can go back to proving

$$\left| \frac{\vec{x} \cdot \vec{y}}{\|\vec{x}\| \, \|\vec{y}\|} \right| \leq 1$$

for $\vec{x}, \vec{y} \neq \vec{0}$.

First, recall one of the properties of absolute value:

The absolute value of a product is the product of the individual absolute values:

$$|ab| = |a| \, |b|$$

Thus,

$$\left| \frac{\vec{x} \cdot \vec{y}}{\|\vec{x}\| \, \|\vec{y}\|} \right| = |\vec{x} \cdot \vec{y}| \left| \frac{1}{\|\vec{x}\| \, \|\vec{y}\|} \right| = \frac{|\vec{x} \cdot \vec{y}|}{\|\vec{x}\| \, \|\vec{y}\|}$$

So now, our condition is just

$$\frac{|\vec{x} \cdot \vec{y}|}{\|\vec{x}\| \, \|\vec{y}\|} \leq 1$$

which is equivalent to

$$|\vec{x} \cdot \vec{y}| \leq \|\vec{x}\| \, \|\vec{y}\|$$

This is known as the *Cauchy-Schwarz inequality*. When expanded, it actually looks like

$$|x_1 y_1 + x_2 y_2 + \ldots + x_n y_n| \leq \sqrt{x_1^2 + x_2^2 + \ldots + x_n^2} \, \sqrt{y_1^2 + y_2^2 + \ldots + y_n^2}$$

But to prove this inequality, we are going to need an idea. In fact,

---

Math Mantra: Cool results can come from solving entirely different problems!

And the problem we are going to look at is one you solved in 9th grade Geometry:

Which point on a given line is closest to the origin?

To solve this problem, you found a perpendicular line that goes through the origin and calculated the intersection:

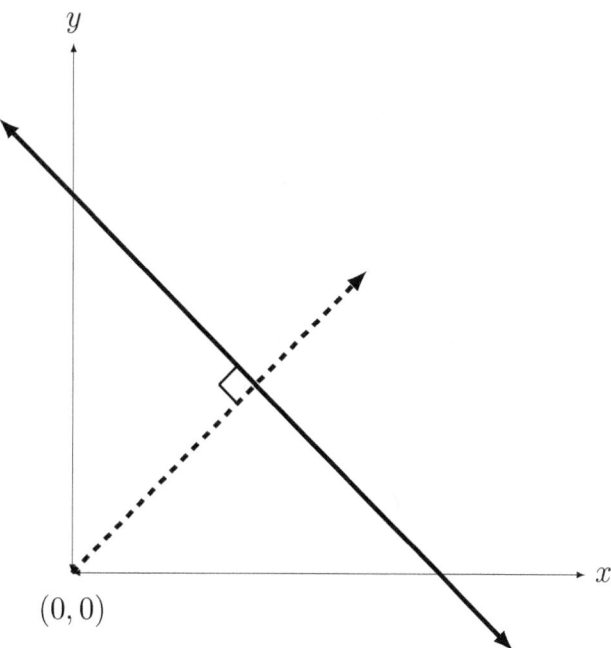

$(0,0)$

However, there is a direct way to solve this problem. Given the line

$$y = mx + b$$

we would like to minimize the distance to the origin:

$$\sqrt{x^2 + y^2} = \sqrt{x^2 + (mx + b)^2}.$$

But what do we do about the square root? The key idea is that this function is minimized precisely when its square is minimized. Precisely,

If we have a non-negative function f, then f and f^2 achieve their minima at the same point x_0.

Why is this true? We showed the reason last lecture:

Square roots preserve inequalities.
and
Squaring both sides preserves inequalities

So if f achieves its minimum at x_0, then

$$f(x_0) \leq f(x)$$

for all x. Squaring,

$$\big(f(x_0)\big)^2 \leq \big(f(x)\big)^2$$

for all x. Hence, f^2 also achieves its minimum at x_0. Conversely, if f^2 achieves its minimum at x_0,

$$\big(f(x_0)\big)^2 \leq \big(f(x)\big)^2$$

for all x. Rooting,

$$f(x_0) \leq f(x)$$

for all x i.e. f also achieves its minimum at x_0.

Thus, we only need to find the x that minimizes

$$
\begin{aligned}
x^2 + (mx+b)^2 &= x^2 + m^2x^2 + 2mbx + b^2 \\
&= \underbrace{(1+m^2)x^2}_{Ax^2} + \underbrace{(2mb)x}_{Bx} + \underbrace{b^2}_{C}
\end{aligned}
$$

But this is a quadratic equation in terms of x. And we all know, from the SAT II or Calc BC, that the minimum of a quadratic $y = Ax^2 + Bx + C$ is achieved at the vertex's x coordinate

$$
x = -\frac{B}{2A}
$$

Here, the minimum is achieved at

$$
x = -\frac{2mb}{2(1+m^2)} = -\frac{mb}{1+m^2}
$$

But this is just for $\mathbb{R}^2$. How do we generalize to higher dimensions?

Again, we cannot visualize higher dimensional lines. Thus, we *define* a line in n-dimensions. Looking at the 2D formula for a line,

$$
y = mx + b,
$$

we make a guess that the equation for the n-dimensional case is

$$
\vec{y} = \vec{m}x + \vec{b}.
$$

This should make intuitive sense: each unit of x increases vector $\vec{y}$ by some $\vec{m}$ "slope." Moreover, just to be clear, a line is the *set* of points satisfying this equation:

$$
\left\{ \vec{y} \in \mathbb{R}^n \mid \vec{y} = \vec{m}x + \vec{b} \text{ for some } x \in \mathbb{R} \right\}
$$

However, we like to reserve x, y for vectors and use t for scalars, so we change variables and use a more common notation:

Definition. *Given vectors $\vec{x}, \vec{y} \in \mathbb{R}^n$, the **line** that goes through $\vec{x}$ in the direction of $\vec{y}$ is the set of all points of the form*

$$
\vec{x} + t\vec{y}
$$

where $t \in \mathbb{R}$.

Now we can prove the n-dimensional analogue of the point on a line closest to the origin:

Lemma. *For $\vec{x}, \vec{y} \in \mathbb{R}^n$ and $\vec{y} \neq \vec{0}$, consider the line that goes through vector $\vec{x}$ in the direction of $\vec{y}$:*

$$\vec{x} + t\vec{y}.$$

Then this line is closest to the origin when

$$t = -\frac{\vec{x} \cdot \vec{y}}{\|\vec{y}\|^2}$$

Proof Summary:

- This is equivalent to minimizing the square of a norm.

- Expand the square into a quadratic using dot product properties

- The minimum of this quadratic occurs at the x coordinate of the vertex.

Proof: We want to find the particular t that minimizes the distance between $\vec{x} + t\vec{y}$ and the origin:

$$\|\vec{x} + t\vec{y}\|.$$

Since this is a non-negative function, we know it achieves its minimum when its square achieves its minimum:

$$\|\vec{x} + t\vec{y}\|^2$$

But now we can use dot product properties:

$$
\begin{aligned}
\|\vec{x} + t\vec{y}\|^2 &= (\vec{x} + t\vec{y}) \cdot (\vec{x} + t\vec{y}) && \text{(Convert Norm to Dot Product)} \\
&= \vec{x} \cdot \vec{x} + \vec{x} \cdot (t\vec{y}) + (t\vec{y}) \cdot \vec{x} + t\vec{y} \cdot t\vec{y} && \text{(Distributive and Commutative Laws)} \\
&= \|\vec{x}\|^2 + 2t(\vec{x} \cdot \vec{y}) + t^2\|\vec{y}\|^2 && \text{(Convert Dot Product to Norm)} \\
&= \underbrace{\|\vec{y}\|^2 t^2}_{At^2} + \underbrace{2(\vec{x} \cdot \vec{y})t}_{Bt} + \underbrace{\|\vec{x}\|^2}_{C}
\end{aligned}
$$

But lo and behold, this is a quadratic equation in terms of t! Using the vertex formula

$$t = -\frac{B}{2A}$$

with

$$
\begin{aligned}
A &= \|\vec{y}\|^2 \\
B &= 2(\vec{x} \cdot \vec{y}),
\end{aligned}
$$

the minimum occurs when

$$t = -\frac{\vec{x} \cdot \vec{y}}{\|\vec{y}\|^2}. \qquad \square$$

Now we use this result to derive one of the most important inequalities in mathematics:

Theorem (Cauchy-Schwarz Inequality). *For any two vectors $\vec{x}, \vec{y} \in \mathbb{R}^n$,*

$$|\vec{x} \cdot \vec{y}| \leq \|\vec{x}\| \, \|\vec{y}\|.$$

Proof Summary:

- We know that $\|\vec{x} + t\vec{y}\|^2$ is always non-negative.

- Plug in the minimizing t from the previous Lemma.

- Isolate the dot product and square root both sides.

Proof: Since a square of a number is always non-negative,

$$\|\vec{x} + t\vec{y}\|^2 \geq 0.$$

From the proof of the lemma, we can rewrite the left side:

$$\underbrace{\|\vec{y}\|^2 t^2 + 2(\vec{x} \cdot \vec{y})t + \|\vec{x}\|^2}_{\|\vec{x}+t\vec{y}\|^2} \geq 0.$$

Now plug in a particular t. Namely, the t where $t\vec{x} + \vec{y}$ is closest to the origin:

$$t = -\frac{\vec{x} \cdot \vec{y}}{\|\vec{y}\|^2}.$$

Then, our inequality becomes

$$\underbrace{\frac{(\vec{x} \cdot \vec{y})^2}{\|\vec{y}\|^2} - 2\frac{(\vec{x} \cdot \vec{y})^2}{\|\vec{y}\|^2} + \|\vec{x}\|^2}_{\|\vec{y}\|^2 t^2 + 2(\vec{x}\cdot\vec{y})t + \|\vec{x}\|^2} \geq 0.$$

Multiplying both sides by $\|\vec{y}\|^2$ and moving terms to the other side, we get

$$\|\vec{y}\|^2 \|\vec{x}\|^2 \geq (\vec{x} \cdot \vec{y})^2.$$

Finally, take the square root of both sides:

$$|\vec{x} \cdot \vec{y}| \leq \|\vec{x}\| \, \|\vec{y}\|.$$

Are we done now? No! To quote Professor Simon,

There is a slight fly in the ointment!

We assumed in our proof $\|\vec{y}\| \neq \vec{0}$ when we defined t, but this may not be the case. Luckily though, the proof is trivial when $\vec{y} = \vec{0}$ since

$$\underbrace{0}_{|\vec{x}\cdot\vec{y}|} \leq \underbrace{0}_{\|\vec{x}\|\,\|\vec{y}\|}.$$

Now we're done! $\square$

Remarkably, we just used a completely different problem (a geometric one) to prove the Cauchy-Schwarz Inequality! That's pretty cool! And because this is so awesome, I'll talk more about this proof technique at the end of this lecture.

Before we move on, take note of two steps in this proof:

1. **We took the square of a norm.**

   Typically,

   Math Mantra: When dealing with norms and square roots, it is often easier
   to work with their squares.

   In particular, when we square a norm, we can expand it as a dot product:

   $$\|\vec{x}\|^2 = \vec{x} \cdot \vec{x}$$

   This allows us to exploit dot product properties.

2. **We took the square root of a square.**

   Near the end of the proof, we calculated

   $$\sqrt{(\vec{x} \cdot \vec{y})^2} = |\vec{x} \cdot \vec{y}|.$$

   Note the **absolute value sign!**

   Generally,
   $$\sqrt{a^2} = |a|$$

   It is **incorrect** to write
   $$\sqrt{a^2} = a.$$

2.4   On Keeping One's Word: Triangle Inequality

As promised in the previous chapter, we will complete our proof that

$$\|\vec{x} - \vec{y}\| \text{ is a distance function.}$$

All that was left to prove was the triangle inequality: for all $\vec{x}, \vec{y}, \vec{z} \in \mathbb{R}^n$,

$$d(\vec{x}, \vec{z}) \leq d(\vec{x}, \vec{y}) + d(\vec{y}, \vec{z})$$

i.e.

$$\|\vec{x} - \vec{z}\| \leq \|\vec{x} - \vec{y}\| + \|\vec{y} - \vec{z}\|.$$

We could prove this directly, but it's a lot easier to prove

$$\|\vec{a} + \vec{b}\| \leq \|\vec{a}\| + \|\vec{b}\|.$$

The result then follows by substituting

$$\vec{a} = \vec{x} - \vec{y}$$
$$\vec{b} = \vec{y} - \vec{z}.$$

Theorem (Triangle Inequality). *For any $\vec{a}, \vec{b} \in \mathbb{R}^n$*

$$\|\vec{a} + \vec{b}\| \le \|\vec{a}\| + \|\vec{b}\|.$$

Proof Summary:

- Expand $\|\vec{a} + \vec{b}\|^2$ using dot product properties.

- Use Cauchy-Schwartz on the $a \cdot b$ term.

- The Right Hand Side (RHS) of the inequality is simply $\left(\|\vec{a}\| + \|\vec{b}\| \right)^2$.

- Square root both sides.

Proof: We once again square the norm and exploit dot product properties:

$$
\begin{aligned}
\|\vec{a} + \vec{b}\|^2 &= \left(\vec{a} + \vec{b} \right) \cdot \left(\vec{a} + \vec{b} \right) \\
&= \vec{a} \cdot \vec{a} + 2 \left(\vec{a} \cdot \vec{b} \right) + \vec{b} \cdot \vec{b} \\
&= \|\vec{a}\|^2 + \|\vec{b}\|^2 + 2 \left(\vec{a} \cdot \vec{b} \right).
\end{aligned}
$$

But $x \le |x|$, so in particular,

$$\vec{a} \cdot \vec{b} \le |\vec{a} \cdot \vec{b}|.$$

Thus, we can create the bound

$$\|\vec{a}\|^2 + \|\vec{b}\|^2 + 2 \left(\vec{a} \cdot \vec{b} \right) \le \|\vec{a}\|^2 + \|\vec{b}\|^2 + 2 \left| \vec{a} \cdot \vec{b} \right|.$$

Using Cauchy-Schwarz, we can bound the right hand side further by

$$\|\vec{a}\|^2 + \|\vec{b}\|^2 + 2|\vec{a} \cdot \vec{b}| \le \|\vec{a}\|^2 + \|\vec{b}\|^2 + 2\|\vec{a}\| \, \|\vec{b}\|.$$

Thus, we have

$$\|\vec{a} + \vec{b}\|^2 \le \|\vec{a}\|^2 + \|\vec{b}\|^2 + 2\|\vec{a}\| \, \|\vec{b}\|$$

Rewriting the right hand side as a square

$$\|\vec{a} + \vec{b}\|^2 \le \left(\|\vec{a}\| + \|\vec{b}\| \right)^2,$$

we root both sides to get

$$\|\vec{a} + \vec{b}\| \le \|\vec{a}\| + \|\vec{b}\|. \qquad \square$$

2.5   Some Fun with Cauchy-Schwarz

There are lots of fun things you can show with Cauchy-Schwarz, but most of the time you'll be using it to establish an upper bound, one typically involving a magic letter named ϵ. But for practice, here are some fun applications:

Example. *For any angle* θ,

$$|\cos^2 \theta - \sin^2 \theta| \leq 1.$$

Proof: Apply the Cauchy-Schwarz inequality to

$$\vec{x} = \begin{bmatrix} \cos \theta \\ \sin \theta \end{bmatrix} \quad \vec{y} = \begin{bmatrix} \cos \theta \\ -\sin \theta \end{bmatrix}$$

Since

$$|\vec{x} \cdot \vec{y}| = |\cos \theta \cos \theta - \sin \theta \sin \theta| = |\cos^2 \theta - \sin^2 \theta|$$

and

$$\|\vec{x}\| \, \|\vec{y}\| = \left(\sqrt{(\cos \theta)^2 + (\sin \theta)^2} \right) \left(\sqrt{(\cos \theta)^2 + (-\sin \theta)^2} \right) = 1,$$

Cauchy-Schwarz tells us

$$\underbrace{|\cos^2 \theta - \sin^2 \theta|}_{|\vec{x} \cdot \vec{y}|} \leq \underbrace{1}_{\|\vec{x}\| \, \|\vec{y}\|} \qquad \square$$

In case you forgot, the left hand side is just the double angle formula for cosine. You used this equation tons of times when integrating $\cos^2 \theta$.

I learned the next Cauchy-Schwarz example from a tenth grade Taiwanese national, who would take the H-series at Stanford 3 years later.

Example. *For any* $a, b, c > 0$,

$$(a + b + c)\left(\frac{1}{a} + \frac{1}{b} + \frac{1}{c} \right) \geq 9$$

Proof: Rewrite the left hand side as the square of a products of norms: let

$$\vec{x} = \begin{bmatrix} \sqrt{a} \\ \sqrt{b} \\ \sqrt{c} \end{bmatrix} \quad \vec{y} = \begin{bmatrix} \dfrac{1}{\sqrt{a}} \\ \dfrac{1}{\sqrt{b}} \\ \dfrac{1}{\sqrt{c}} \end{bmatrix}$$

Then,

$$\left(\|\vec{x}\|\,\|\vec{y}\|\right)^2 = \|\vec{x}\|^2\,\|\vec{y}\|^2 = (a+b+c)\left(\frac{1}{a}+\frac{1}{b}+\frac{1}{c}\right)$$

We also know

$$|\vec{x}\cdot\vec{y}|^2 = (1+1+1)^2 = 9$$

So, by the square of Cauchy-Schwarz,

$$\underbrace{9}_{|\vec{x}\cdot\vec{y}|^2} \le \underbrace{(a+b+c)\left(\frac{1}{a}+\frac{1}{b}+\frac{1}{c}\right)}_{(\|\vec{x}\|\,\|\vec{y}\|)^2} \qquad\qquad \square$$

The next example is a famous one that translates to

The square of the average is less than or equal to the average of the squares.

If you are not sure what this means, it is always good to write down an example:

$$\left(\frac{1+2+5+6+10}{5}\right)^2 \le \frac{1^2+2^2+5^2+6^2+10^2}{5}$$

Example. *Let $a_1, a_2, \ldots, a_n \in \mathbb{R}$. Then,*

$$\left(\frac{1}{n}\sum_{i=1}^{n} a_i\right)^2 \le \frac{1}{n}\sum_{i=1}^{n} a_i^2$$

Proof: The magic vectors we are going to use this time are

$$\vec{x} = \begin{bmatrix} a_1 \\ a_2 \\ \vdots \\ a_n \end{bmatrix} \qquad \vec{y} = \begin{bmatrix} \dfrac{1}{n} \\ \dfrac{1}{n} \\ \vdots \\ \dfrac{1}{n} \end{bmatrix}$$

First, calculate the dot product as

$$|\vec{x}\cdot\vec{y}| = \sum_{i=1}^{n}\frac{a_i}{n} = \frac{1}{n}\sum_{i=1}^{n} a_i$$

and the norm product as

$$\|\vec{x}\|\,\|\vec{y}\| = \sqrt{\sum_{i=1}^{n} a_i^2}\sqrt{\sum_{i=1}^{n}\frac{1}{n^2}} = \frac{1}{\sqrt{n}}\sqrt{\sum_{i=1}^{n} a_i^2}.$$

So Cauchy-Schwarz tells us

$$\underbrace{\frac{1}{n}\sum_{i=1}^{n}a_i}_{|\vec{x}\cdot\vec{y}|} \leq \underbrace{\frac{1}{\sqrt{n}}\sqrt{\sum_{i=1}^{n}a_i^2}}_{\|\vec{x}\|\,\|\vec{y}\|}.$$

Since we have non-negative terms, we can square both sides:

$$\left(\frac{1}{n}\sum_{i=1}^{n}a_i\right)^2 \leq \frac{1}{n}\sum_{i=1}^{n}a_i^2. \qquad \square$$

The final example doesn't have a cute saying (at least not one I can think of). But the symmetry is pretty sweet, and just like in love, it is easier to see beauty than describe it.

Example. *Let $a_1, a_2, \ldots, a_n$ be positive. Then,*

$$a_1 + a_2 + \ldots + a_n \leq \frac{a_1^2}{a_2} + \frac{a_2^2}{a_3} + \ldots + \frac{a_{n-1}^2}{a_n} + \frac{a_n^2}{a_1}$$

Proof: We will apply Cauchy-Schwarz to the vectors

$$\vec{x} = \begin{bmatrix} \sqrt{a_2} \\ \sqrt{a_3} \\ \vdots \\ \sqrt{a_n} \\ \sqrt{a_1} \end{bmatrix} \qquad \vec{y} = \begin{bmatrix} \dfrac{a_1}{\sqrt{a_2}} \\ \dfrac{a_2}{\sqrt{a_3}} \\ \vdots \\ \dfrac{a_{n-1}}{\sqrt{a_n}} \\ \dfrac{a_n}{\sqrt{a_1}} \end{bmatrix}$$

Then,

$$|\vec{x}\cdot\vec{y}| = \left|\sqrt{a_2}\frac{a_1}{\sqrt{a_2}} + \sqrt{a_3}\frac{a_2}{\sqrt{a_3}} + \ldots + \sqrt{a_n}\frac{a_{n-1}}{\sqrt{a_n}} + \sqrt{a_1}\frac{a_n}{\sqrt{a_1}}\right| = |a_1 + a_2 + \ldots + a_n|.$$

But we can drop the absolute value since the a_i are non-negative:

$$|\vec{x}\cdot\vec{y}| = a_1 + a_2 + \ldots + a_n.$$

Also,

$$\|\vec{x}\|\,\|\vec{y}\| = \sqrt{(\sqrt{a_2})^2 + (\sqrt{a_3})^2 + \ldots + (\sqrt{a_n})^2 + (\sqrt{a_1})^2}\sqrt{\left(\frac{a_1}{\sqrt{a_2}}\right)^2 + \left(\frac{a_2}{\sqrt{a_3}}\right)^2 + \ldots + \left(\frac{a_{n-1}}{\sqrt{a_n}}\right)^2 + \left(\frac{a_n}{\sqrt{a_1}}\right)^2}$$

$$= \sqrt{a_2 + a_3 + \ldots + a_n + a_1}\sqrt{\frac{a_1^2}{a_2} + \frac{a_2^2}{a_3} + \ldots + \frac{a_{n-1}^2}{a_n} + \frac{a_n^2}{a_1}}$$

So by Cauchy-Schwarz,

$$\underbrace{a_1 + a_2 + \ldots + a_n}_{|\vec{x} \cdot \vec{y}|} \leq \underbrace{\sqrt{a_2 + a_3 + \ldots + a_n + a_1}\sqrt{\frac{a_1^2}{a_2} + \frac{a_2^2}{a_3} + \ldots + \frac{a_{n-1}^2}{a_n} + \frac{a_n^2}{a_1}}}_{\|\vec{x}\|\,\|\vec{y}\|}$$

Since both sides are non-negative, we can square this inequality to get

$$(a_1 + a_2 + \ldots + a_n)^2 \leq (a_2 + a_3 + \ldots + a_n + a_1)\left(\frac{a_1^2}{a_2} + \frac{a_2^2}{a_3} + \ldots + \frac{a_{n-1}^2}{a_n} + \frac{a_n^2}{a_1}\right).$$

Notice that

$$a_1 + a_2 + \ldots + a_n = a_2 + a_3 + \ldots + a_n + a_1.$$

Thus, we can divide out by

$$a_2 + a_3 + \ldots + a_n + a_1$$

to get

$$a_1 + a_2 + \ldots + a_n \leq \frac{a_1^2}{a_2} + \frac{a_2^2}{a_3} + \ldots + \frac{a_n^2}{a_1}. \qquad \square$$

After seeing these examples, the natural question to ask is

> *When is equality achieved in the Cauchy-Schwarz inequality?*

That's a very good question! We will save this discussion for the next chapter, after we introduce *if and only if* proofs.

One last remark: in the preceding proofs, notice that I did not tell you *how* we cooked up $\vec{x}$ and $\vec{y}$. This is because:

> Math Mantra: Some things are not obvious and you really have to mull them over.

And often times, this will involve tons of scratch work. But that is what makes mathematics an art.

2.6 Proof Technique: The 7-10 Split

The hardest shot in bowling is the *7-10 Split*:

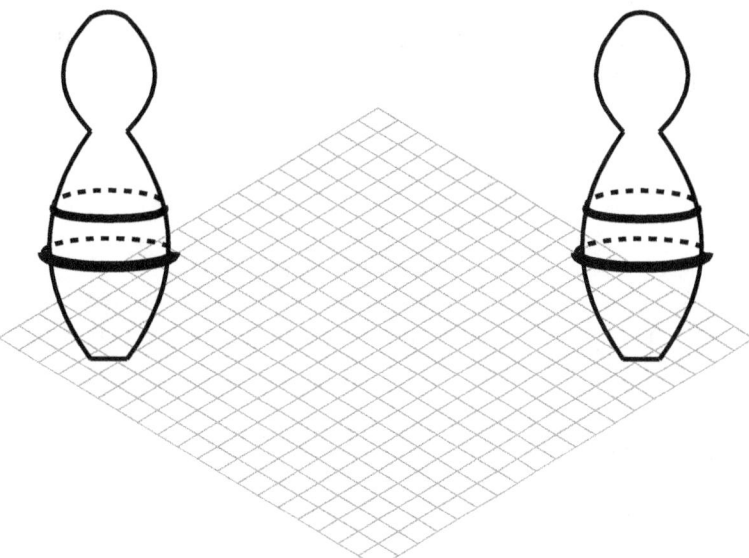

To knock down both pins, the ball has to collide in such a way that one pin ricochets into the other. Amazingly, by trying to knock down one pin, it leads to the collapse of the other.

In math, we build theorems that often have immediate implications. But some of the rarest and most beautiful proofs are the ones that pop out of nowhere through the solution of a *seemingly unrelated* problem.

I hesitate mentioning this proof technique now because

- You haven't learned all the basic proof techniques.

- Describing this proof technique is the mathematical equivalent of trying to explain *irony*.

But this is indeed the trick that we used to prove Cauchy-Schwarz: we derived this inequality by solving the geometric problem of finding the point on a given line that is closest to the origin.

There are some fantastic examples: my personal favorite is Prufer codes.[1] A Prufer code solves the problem of storing data structures known as trees. But we can also use it to solve a seemingly unrelated problem: the problem of counting the number of labelled trees with n nodes.

---

[1]Wikipedia it! The math is very simple and will be especially useful if you have CS aspirations. You will also see Prufer codes in **Math 108.**

The prototypical[2] example of the 7-10 split is one you will prove in 52H. Recall, in Calc BC, you used *p*-series to prove that

$$\sum_{n=1}^{\infty} \frac{1}{n^2}$$

converges. But what does it actually converge to? Using the entirely different subject of Fourier Series, the answer magically pops out:

$$\sum_{n=1}^{\infty} \frac{1}{n^2} = \frac{\pi^2}{6}.$$

Awesome!

However, these examples involve words you've never seen before (like trees and Fourier Series). But you came here to do math! You want to see theorems that you can **understand**. So here is an example from your Algebra II days.

The Binomial Theorem states that

$$(x+y)^n = \sum_{i=0}^{n} \binom{n}{i} x^i y^{n-i}$$

This tells us how to expand a power of a binomial. But we can actually use this to prove set theoretic facts!

We say that

a set A is a ***subset*** of a set B if for every x in A, x is also in B.

Often, this expression will be denoted by

$$A \subseteq B$$

Alternatively, we can write

$$B \supseteq A$$

The first property is that any set of size n has 2^n subsets. For example, the 3 element set

$$\{0, 1, 2\}$$

has 2^3 subsets, namely,

$$\{\}$$
$$\{0\} \qquad \{1\} \qquad \{2\}$$
$$\{0, 1\} \quad \{0, 2\} \quad \{1, 2\}$$
$$\{0, 1, 2\}$$

where $\{\}$ is just the empty set.

Now we prove this property using the Binomial Theorem:

---

[2]The prototypical example is actually elliptic curves and Fermat's Last Theorem, but I make it a policy to never talk about anything I can't prove. Guess I'd make a pretty lousy politician :)

Example. *Any set of size n has 2^n subsets.*

Proof Summary:

- Expand $(1+1)^n$ using the Binomial Theorem

- Interpret the right hand side as the number of subsets.

Proof: Plugging $x = 1$ and $y = 1$ into the Binomial Theorem yields

$$(1+1)^n = \sum_{i=0}^{n} \binom{n}{i} 1^i 1^{n-i}$$

which simplifies to

$$2^n = \sum_{i=0}^{n} \binom{n}{i}$$

But how do we build all the subsets of an n-element set? The number of ways to build a subset of size i is simply the number of ways to choose i out of the original n elements:

$$\binom{n}{i}$$

So the number of subsets is

$$\underbrace{\binom{n}{0}}_{\text{Number of 0-element subsets}} + \underbrace{\binom{n}{1}}_{\text{Number of 1-element subsets}} + \cdots + \underbrace{\binom{n}{n}}_{\text{Number of n-element subsets}} .$$

which is the right hand side above, so it equals

$$2^n$$

by the Binomial Theorem.   □

We can use the same trick to prove that any set has an equal number of odd and even sized subsets. In our previous example, $\{0, 1, 2\}$, the number of even sized subsets is 4:

$$\{\}$$
$$\{0, 1\} \quad \{0, 2\} \quad \{1, 2\}$$

whereas the number of odd size subsets is also 4:

$$\{0\} \quad \{1\} \quad \{2\}$$
$$\{0, 1, 2\}$$

Example. *For any set of size n, the number of odd sized subsets is equal to the number of even sized subsets.*

Proof Summary:

- Expand $(-1+1)^n$ using the Binomial Theorem.

- Move all negative $\binom{n}{i}$ terms to the left side of the equation.

- Interpret the left side as the number of even subsets and the right side as the number of odd subsets.

Proof: For ease, let's assume n is even (the argument for n odd is essentially the same). Plugging $x = -1$ and $y = 1$ into the Binomial Theorem yields

$$(-1+1)^n = \sum_{i=0}^{n} \binom{n}{i}(-1)^i(1)^{n-i}$$

which simplifies to

$$0 = \binom{n}{0} - \binom{n}{1} + \binom{n}{2} - \binom{n}{3} + \ldots + \binom{n}{n}.$$

Moving all the negative terms to the left,

$$\binom{n}{1} + \binom{n}{3} + \binom{n}{5} + \ldots + \binom{n}{n-1} = \binom{n}{0} + \binom{n}{2} + \binom{n}{4} + \ldots + \binom{n}{n}.$$

But this just says the number of odd size subsets is the same as the number of even size subsets. □

<u>**New Notation**</u>

Symbol	Reading	Example	Example Translation
$\vec{x} \cdot \vec{y}$	The dot product of vectors $\vec{x}$ and $\vec{y}$	$\vec{x} \cdot \vec{y} < 9$	The dot product of vectors $\vec{x}$ and $\vec{y}$ is less than 9.
(LHS)	The left hand side	The (LHS) of $A = B$ is A	The left hand side of equation $A = B$ is A.
(RHS)	The right hand side	The (RHS) of $A = B$ is B	The right hand side of equation $A = B$ is B.
$A \subseteq B$	A is a subset of B	$\mathbb{Z} \subseteq \mathbb{Q}$	The set of integers is contained in the set of rationals.
$B \supseteq A$	B contains the subset A	$\mathbb{C} \supseteq \mathbb{R}$	The set of complex numbers contains the set of reals.

Lecture 3

Let's get Linear!

Its been quite a thousand years, and I feel some obligation to contribute to the millennial reminiscences before we get too far into the new century. I propose linearity as one of the most important themes of mathematics and its applications in times past and present.

-The Great Brad Osgood, *EE261*.

Goals: First, we define linear functions and use them to motivate the definition of a subspace (of $\mathbb{R}^n$). Afterwards, we introduce the notion of span to construct subspaces from some initial set of vectors. In this pursuit, natural questions arise concerning redundancy of vectors, allowing us to introduce Proof by Contradiction and if and only if statements.

3.1 Linear Functions

For the first few days, we have been studying "Linear Algebra." But what exactly makes it *Linear*?

Linearity is a property that is pretty much everywhere, from physics to finance. To explain linearity, let's consider a simple example:

Suppose

x *ounces of rum produces* $f(x)$ *Mojitos.*

Intuitively, if you double the input, you should double the output:

$$f(2x) = 2f(x)$$

Likewise, if

x *ounces of rum produces* $f(x)$ *Mojitos.*
y *ounces of rum produces* $f(y)$ *Mojitos.*

Then, $x + y$ ounces of rum should produce $f(x) + f(y)$ Mojitos:

$$f(x + y) = f(x) + f(y).$$

Simple, right? Precisely,

Definition. *Let f be a function with domain V. We say that f is **linear** on V if it satisfies two properties:*

1. *(**Superposition**) Applying a function to a sum of inputs is the same as summing the function applied to each term separately:*

$$f(x + y) = f(x) + f(y)$$

for any $x, y \in V$.

2. *(**Homogeneity**) Scaling the input by α scales the output by α:*

$$f(\alpha x) = \alpha f(x)$$

for any $x \in V$ and real number α.

Next week, you will see that the most important linear functions in this course involve matrix multiplication in $\mathbb{R}^n$, namely:

$$f(x) = A\vec{x}$$

where A is an $n \times n$ matrix. In fact, you will learn that *any* linear function on $\mathbb{R}^n$ **can be expressed as a matrix multiplication!**

3.2   Subspaces of $\mathbb{R}^n$

Notice that I cheated in the definition of a linear function: I never defined its domain and just left it as V. If you take **Math 113**, **Math 171**, and **Math 121**, you will learn that V is an abstract collection known as a *vector space.*

For example, suppose

V is the set of differentiable functions.

Then the derivative is a linear function on V: the derivative of the sum of two functions is just the sum of the individual derivatives

$$\frac{d}{dx}\left[f(x) + g(x)\right] = \frac{d}{dx}f(x) + \frac{d}{dx}g(x)$$

and you can always pull a scalar out of a derivative.

$$\frac{d}{dx}\left[\alpha f(x)\right] = \alpha \frac{d}{dx}f(x)$$

Likewise, if

V is the set of integrable functions,

the integral is linear on V:

$$\int \left[f(x) + g(x) \right] dx = \int f(x)\, dx + \int g(x)\, dx$$

$$\int \left[\alpha f(x) \right] dx = \alpha \int f(x)\, dx$$

But, **for this course**, we are going to assume

$$V \text{ is a subset of } \mathbb{R}^n.$$

Moreover, we require V to have certain properties.

Recall that our definition of a linear function took V to be the domain of f. The property

$$f(x) + f(y) = f(x + y)$$

only makes sense if the domain is closed under addition: if x and y are in V, then so is $x + y$.

Also,

$$\alpha f(x) = f(\alpha x)$$

only makes sense if the domain is closed under scaling: if x is in the V, so is αx.

This motivates us to make the following definition for such a domain V:

Definition. *A **subspace** is a subset $V \subseteq \mathbb{R}^n$ that satisfies the following properties:*

1. *(**Existence of Zero**) The zero vector is in V:*

$$\vec{0} \in V$$

2. *(**Closure under Addition**) Given two vectors in V, their sum is also in V:*

$$\vec{x} + \vec{y} \in V$$

for all $\vec{x}, \vec{y} \in V$

3. *(**Closure under Scalar Multiplication**) For any vector, all scalar multiples of that vector are in V:*

$$\alpha \vec{x} \in V$$

for all $\vec{x} \in V$ and real α.

Note that we added the *Existence of Zero* requirement so that the empty set is not a subspace.

3.3   How to Verify a Set is a Subspace

I guarantee a significant percentage of students who scored a 5 on the Calculus BC
cannot prove the span is a subspace.

-Leon Simon

Professor Simon's assertion is **absolutely correct**. The reason is that students have yet to learn how to prove a *universal* statement. However, after you've mastered this technique, you'll realize that verifying that a set is a subspace is straightforward. Just check the definition!

Example. *The set*

$$A = \left\{ \begin{bmatrix} x \\ y \\ z \end{bmatrix} \in \mathbb{R}^3 \,\middle|\, x = y = z \right\}$$

is a subspace.

Proof: We need to directly check the definition of a subspace.

- *Existence of Zero*
  Immediately

$$\begin{bmatrix} 0 \\ 0 \\ 0 \end{bmatrix} \in A$$

  since its components are equal.

- *Closure under Addition*
  Consider any two vectors $\vec{x}, \vec{y} \in A$. By definition, they are of the form

$$\vec{x} = \begin{bmatrix} c \\ c \\ c \end{bmatrix}$$

$$\vec{y} = \begin{bmatrix} d \\ d \\ d \end{bmatrix}$$

  for some reals c, d. Then,

$$\vec{x} + \vec{y} = \begin{bmatrix} c \\ c \\ c \end{bmatrix} + \begin{bmatrix} d \\ d \\ d \end{bmatrix} = \begin{bmatrix} c+d \\ c+d \\ c+d \end{bmatrix}$$

  Each of its components are equal; therefore,

$$\vec{x} + \vec{y} \in A.$$

- *Closure under Scalar Multiplication*
  Consider any vector $\vec{x} \in A$. By definition,

$$\vec{x} = \begin{bmatrix} c \\ c \\ c \end{bmatrix}$$

for some constant c. Then, for any real k,

$$k\vec{x} = k\begin{bmatrix} c \\ c \\ c \end{bmatrix} = \begin{bmatrix} kc \\ kc \\ kc \end{bmatrix}$$

which again has equal components, so
$$k\vec{x} \in A. \qquad \square$$

You also need to know how to prove theorems given *arbitrary* subspaces.

We define

The **intersection** of A, B, denoted by $A \cap B$, is the set of all elements in both A **and** B.

Using this notation, we can prove

Example. *For any subspaces $A, B \subseteq \mathbb{R}^n$,*
$$A \cap B$$
is a subspace.

Proof:

- *Existence of Zero*
  By definition of a subspace,
$$\vec{0} \in A$$
$$\vec{0} \in B$$

  Therefore,
$$\vec{0} \in A \cap B.$$

- *Closure under Addition*
  Consider any two vectors $\vec{x}, \vec{y} \in A \cap B$. This means
$$\vec{x} \in A$$
$$\vec{y} \in A$$

  Since A is a subspace,
$$\vec{x} + \vec{y} \in A$$

Likewise,

$$\vec{x} \in B$$
$$\vec{y} \in B$$

and thus,

$$\vec{x} + \vec{y} \in B$$

Therefore,

$$\vec{x} + \vec{y} \in A \cap B.$$

- *Closure under Scalar Multiplication*
  Consider any vector $\vec{x} \in A \cap B$. Then,

$$\vec{x} \in A$$
$$\vec{x} \in B$$

and so for any real k,

$$k\vec{x} \in A$$
$$k\vec{x} \in B.$$

Therefore,

$$k\vec{x} \in A \cap B.$$ $\square$

The last two examples will be of fundamental importance in **Lecture 8**:

Example. *Let* $f : \mathbb{R}^n \to \mathbb{R}^m$ *be linear and let* $V \subseteq \mathbb{R}^n$ *be a subspace. The image of* V *under* f,

$$f(V) = \{ f(\vec{x}) \mid \vec{x} \in V \}$$

is a subspace.

Proof:

- *Existence of Zero*
  By definition of a subspace,

$$\vec{0} \in V$$

Applying linearity of f,

$$f(\vec{0}) = f(\vec{0} + \vec{0}) = f(\vec{0}) + f(\vec{0}).$$

This gives us

$$f(\vec{0}) = \vec{0}$$

so

$$\underbrace{\vec{0}}_{f(\vec{0})} \in f(V).$$

- *Closure under Addition*
  Consider any two vectors $\vec{x}, \vec{y} \in f(V)$. By definition of $f(V)$, there exist vectors $\vec{v}_1, \vec{v}_2 \in V$ such that
  $$f(\vec{v}_1) = \vec{x}$$
  $$f(\vec{v}_2) = \vec{y}$$
  Closure of V (under addition) yields
  $$\vec{v}_1 + \vec{v}_2 \in V.$$
  This gives us
  $$f(\vec{v}_1 + \vec{v}_2) \in f(V).$$
  Applying linearity,
  $$\underbrace{f(\vec{v}_1) + f(\vec{v}_2)}_{f(\vec{v}_1 + \vec{v}_2)} \in f(V).$$
  i.e.
  $$\vec{x} + \vec{y} \in f(V).$$

- *Closure under Scalar Multiplication*
  Consider any vector $\vec{x} \in f(V)$. Then,
  $$f(\vec{v}) = \vec{x}$$
  for some $\vec{v} \in V$. Since V is a subspace, for any real k,
  $$k\vec{v} \in V.$$
  Now we have
  $$f(k\vec{v}) \in f(V).$$
  Applying linearity,
  $$\underbrace{kf(\vec{v})}_{f(k\vec{v})} \in f(V)$$
  i.e
  $$k\vec{x} \in f(V). \qquad \square$$

Example. *Let $f : \mathbb{R}^n \to \mathbb{R}^m$ be linear and let $V \subseteq \mathbb{R}^n$ be a subspace. The solution set of $f(\vec{x}) = \vec{0}$,*
$$N = \left\{ \vec{x} \in \mathbb{R}^n \,\middle|\, f(\vec{x}) = \vec{0} \right\}$$
is a subspace.

Proof:

- *Existence of Zero*
  Applying linearity of f,
  $$f(\vec{0}) = f(\vec{0} + \vec{0}) = f(\vec{0}) + f(\vec{0}).$$

This gives us

$$f(\vec{0}) = \vec{0}.$$

Therefore,

$$\vec{0} \in N.$$

- *Closure under Addition*
  Consider any two vectors $\vec{x}, \vec{y} \in N$. By definition, they satisfy

$$
\begin{aligned}
f(\vec{x}) &= \vec{0} \\
f(\vec{y}) &= \vec{0}
\end{aligned}
$$

By linearity,

$$f(\vec{x} + \vec{y}) = \underbrace{f(\vec{x})}_{\vec{0}} + \underbrace{f(\vec{y})}_{\vec{0}} = \vec{0}.$$

In other words,

$$\vec{x} + \vec{y} \in N.$$

- *Closure under Scalar Multiplication*
  Consider any vector $\vec{x} \in N$. Then,

$$f(\vec{x}) = \vec{0}.$$

For any real k,

$$f(k\vec{x}) = k \underbrace{f(\vec{x})}_{\vec{0}} = \vec{0}$$

by linearity. Therefore,

$$k\vec{x} \in N. \qquad \square$$

3.4   Spanning Vectors

Now that we know what subspaces are, how do we *build* them?

First, we start with some initial set of vectors, say

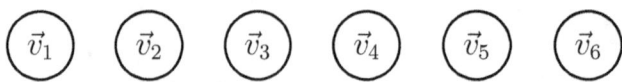

Then, from this initial set of vectors, we are going to *grow* a full subspace S:

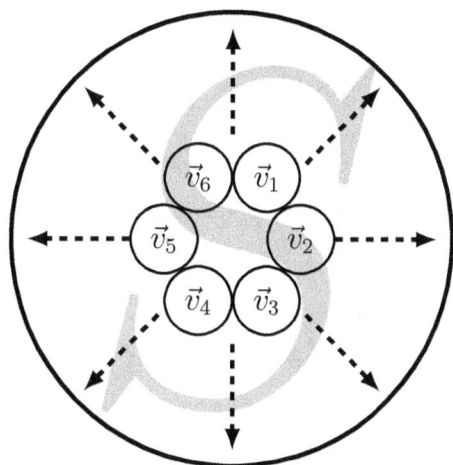

Formally, we take S to be the set of all linear combinations of our initial vectors:

Definition. *A **linear combination** of $\vec{v}_1, \vec{v}_2, \ldots, \vec{v}_k \in \mathbb{R}^n$ is a vector of the form*

$$c_1 \vec{v}_1 + c_2 \vec{v}_2 + \ldots + c_k \vec{v}_k$$

*where $c_1, c_2, \ldots, c_k \in \mathbb{R}$. The **span** of $\vec{v}_1, \vec{v}_2, \ldots, \vec{v}_k$ is the set of all such linear combinations:*

$$\text{span}\{\vec{v}_1, \vec{v}_2, \ldots, \vec{v}_k\} = \{c_1 \vec{v}_1 + c_2 \vec{v}_2 + \ldots + c_k \vec{v}_k \mid c_1, c_2, \ldots, c_k \in \mathbb{R}\}$$

Intuitively, we know the span is a subspace: it is the collection of *all possible sums and scalings* from an initial set of vectors. Of course, we need a proof:

Theorem. *For any vectors $\vec{v}_1, \vec{v}_2, \ldots, \vec{v}_k \in \mathbb{R}^n$,*

$$\text{span}\{\vec{v}_1, \vec{v}_2, \ldots, \vec{v}_k\}$$

is a subspace.

Proof:

- *Existence of Zero*
  Consider a linear combination with each $c_i = 0$:

$$0\vec{v}_1 + 0\vec{v}_2 + \ldots + 0\vec{v}_k = \vec{0}.$$

Thus,

$$\vec{0} \in \text{span}\{\vec{v}_1, \vec{v}_2, \ldots, \vec{v}_k\}.$$

- *Closure under Addition*
  Let $\vec{x}, \vec{y} \in \text{span}\{\vec{v}_1, \vec{v}_2, \ldots, \vec{v}_k\}$. By definition of span,

$$\vec{x} = c_1\vec{v}_1 + c_2\vec{v}_2 + \ldots + c_k\vec{v}_k \quad \text{for some } c_1, c_2, \ldots, c_k \in \mathbb{R}$$
$$\vec{y} = d_1\vec{v}_1 + d_2\vec{v}_2 + \ldots + d_k\vec{v}_k \quad \text{for some } d_1, d_2, \ldots, d_k \in \mathbb{R}$$

Then,

$$\vec{x} + \vec{y} = (c_1 + d_1)\vec{v}_1 + (c_2 + d_2)\vec{v}_2 + \ldots + (c_k + d_k)\vec{v}_k,$$

which is still a linear combination of $\vec{v}_1, \vec{v}_2, \ldots, \vec{v}_k$. Hence,

$$\vec{x} + \vec{y} \in \text{span}\{\vec{v}_1, \vec{v}_2, \ldots, \vec{v}_k\}.$$

- *Closure under scaling*
  For all $\vec{x} \in \text{span}\{\vec{v}_1, \vec{v}_2, \ldots, \vec{v}_k\}$,

$$\vec{x} = c_1\vec{v}_1 + c_2\vec{v}_2 + \ldots + c_k\vec{v}_k \text{ for some } c_1, c_2, \ldots, c_k \in \mathbb{R}.$$

Then for any real α,

$$\alpha\vec{x} = \alpha(c_1\vec{v}_1 + c_2\vec{v}_2 + \ldots + c_k\vec{v}_k) = (\alpha c_1)\vec{v}_1 + (\alpha c_2)\vec{v}_2 + \ldots + (\alpha c_k)\vec{v}_k$$

which is again a linear combination in our span. Thus,

$$\alpha\vec{x} \in \text{span}\{\vec{v}_1, \vec{v}_2, \ldots, \vec{v}_k\}. \qquad \square$$

At this point, you should be asking yourself:

- Can all subspaces be built from a span?

- Can you write one subspace as two different spans?

- Are all the vectors in our spanning list necessary?

The answer to the first question, remarkably, is yes. We will prove this fact in **Lecture 6**.

The answer to the second question is also yes:

$$\text{span}\left\{\begin{bmatrix} 1 \\ 1 \end{bmatrix}, \begin{bmatrix} 1 \\ 0 \end{bmatrix}\right\} \quad \text{and} \quad \text{span}\left\{\begin{bmatrix} 1 \\ 0 \end{bmatrix}, \begin{bmatrix} 0 \\ 1 \end{bmatrix}\right\}$$

are two different ways of writing $\mathbb{R}^2$. In fact, we will prove that we can find a "best" set in **Lecture 31**.

As for the third question, the answer is: **not always**. Consider the two spans,

$$\text{span}\left\{\begin{bmatrix} 1 \\ 1 \\ 1 \end{bmatrix}, \begin{bmatrix} 1 \\ 1 \\ 0 \end{bmatrix}\right\} \quad \text{span}\left\{\begin{bmatrix} 1 \\ 1 \\ 1 \end{bmatrix}, \begin{bmatrix} 1 \\ 1 \\ 0 \end{bmatrix}, \begin{bmatrix} 3 \\ 3 \\ 2 \end{bmatrix}\right\}$$

These two spans are the *same* set. This is because the extra vector in the second list is redundant:

$$\begin{bmatrix} 3 \\ 3 \\ 2 \end{bmatrix} = 2 \begin{bmatrix} 1 \\ 1 \\ 1 \end{bmatrix} + \begin{bmatrix} 1 \\ 1 \\ 0 \end{bmatrix},$$

Formally, we say that the list is *linearly dependent*.

Before we can go into linear dependence, we are going to describe a very powerful proof technique called *Proof by Contradiction*. This will be useful not only in proving results about linear independence, but in **many** proofs throughout your mathematical career. Next, we will clarify, rigorously, the meaning of "equivalent" by showing how to prove *if and only if* statements. Finally, we will apply our new techniques to prove that showing linear independence of a set of vectors is equivalent to showing that none of these vectors can be written as a linear combination of the others.

3.5 Proof Technique: Proof by Contradiction

Mathematicians are strange creatures, one may observe: They go into long arguments based on assumptions they know are false, and their happiest moments are when they find a contradiction between statements they have proved.

-L. Lovasz, *Discrete Mathematics*

One of the basic mathematical axioms is the law of excluded middle:

For any statement P, either P is true or P is not true.

In case you want to get it tattooed, the symbolic shorthand is

$$P \vee \neg P$$

In plain English, this means that for any event, it is always the case that it either happens or it does not happen.

For example, the following are always true:[1]

- I left my glasses in the library or I did not leave my glasses in the library

- You beat me at chess or you didn't beat me at chess.

- Carly Rae Jepsen will call or Carly Rae Jepsen will not call.

In each case, exactly one of the two possibilities is always true: they cannot both be true (by definition of not) and at least one must be true by our axiom.

Why is this useful? We can assume one choice. If the assumption of this choice leads to a contradiction, we know the other choice must have been true! This approach is known as *Proof by Contradiction*.

There are many reasons why you should use *Proof by Contradiction*:

---

[1]Such statements that are always true are called tautologies.

- Proof by contradiction gives us a **jumping point**:

> ```
> Math Mantra:   Assuming the negation of the statement you are trying to prove
>     gives you extra information to work with and can start you in the right
>                                direction.
> ```

- Proof by contradiction is often the **natural way** to approach a problem:

> ```
> Math Mantra:   To prove a definition that is the negation of another, it is
>                  natural to use proof by contradiction.
> ```

For example, irrational is defined as **not** rational, disconnected means **not** connected, and in our case, linearly independent means **not** linearly dependent. In these cases, it is very natural to use proof by contradiction!

Proof by contradiction is very beautiful and is perhaps the most useful proof technique you will use during your undergraduate career. In the words of my Math 171 TA, David Sher:

*Start every proof with **suppose not**!*

As usual with new proof techniques, I give a few examples:

Example. *The product of an irrational number and a non-zero rational number is always irrational.*

Proof: Suppose not. Then there exists an irrational number x and non-zero rational y whose product is rational:

$$xy = \frac{p_1}{q_1}$$

for some integers p_1, q_1. Since y is rational, by definition,

$$y = \frac{p_2}{q_2}$$

for some integers p_2, q_2. Plugging y back into the original equation, we get

$$x \cdot \underbrace{\frac{p_2}{q_2}}_{y} = \frac{p_1}{q_1}.$$

Thus,

$$x = \frac{p_1 q_2}{q_1 p_2}.$$

But this means that x is rational, a contradiction. In conclusion, a product of an irrational and a non-zero rational is always irrational. $\qquad\square$

One of the fundamental facts about the integers is that any integer $n \geq 2$ can be written as a product of primes:

$$n = p_1 p_2 \ldots p_n.$$

For example,

$$360 = 2 \cdot 2 \cdot 2 \cdot 3 \cdot 3 \cdot 5.$$

We will use this fact to prove the following result:

Example. *If n is composite, then n is divisible by some prime less than or equal to $\sqrt{n}$.*

Proof Summary:

- Suppose not: n's prime factors are all greater than $\sqrt{n}$.

- n has at least 2 primes in its factorization (not necessarily distinct).

- Conclude $n > n$, contradiction.

Proof: Suppose not. Then there is some composite n whose prime factors are all greater than $\sqrt{n}$. Moreover, we also know that n has at least two prime factors in its prime factorization (otherwise n would be prime):

$$n = p_1 p_2 \ldots p_k$$

for $k \geq 2$. In particular, n must be greater than (or equal to) the product of its first two prime factors:

$$n \geq p_1 p_2$$

But by our assumption on n,

$$p_1 p_2 > \sqrt{n}\sqrt{n} = n.$$

This gives us

$$n > n$$

which is impossible. In conclusion, every composite number n is divisible by some prime less than or equal to $\sqrt{n}$. $\square$

Even though this is a fast proof, it gives us a neat way to check if a number is prime. For example, to prove

$$499 \text{ is prime}$$

we just need to check that 499 it is not divisible by

$$2, 3, 5, 7, 11, 13, 17, 19.$$

The next example is the classic result that $\sqrt{2}$ is irrational:

Example. $\sqrt{2}$ *is irrational.*

Proof Summary:

- Suppose $\sqrt{2}$ is rational with $\gcd(a, b) = 1$.

- Show b is even.

- Substitute $b = 2k$ and show a is even.

- Contradict $\gcd(a, b) = 1$.

Proof: Suppose $\sqrt{2}$ is rational. Then,

$$\sqrt{2} = \frac{a}{b}$$

where we can assume that a, b are integers such that $\gcd(a, b) = 1$ (we can always divide out common factors). Then, by isolating a and squaring both sides, we have

$$2b^2 = a^2$$

So a^2 is even. This implies a must be even: otherwise if a were odd, a^2 is odd, which is not the case. Thus,

$$a = 2n$$

for some integer n. Then, substituting this back into the equation,

$$2b^2 = \underbrace{(2n)^2}_{a^2}.$$

Simplifying, we get

$$b^2 = 2n^2.$$

So b^2 is even, and thus b is even. But this means that a and b are both divisible by 2, contradicting $\gcd(a, b) = 1$. Thus we have a contradiction and $\sqrt{2}$ is irrational. □

Next, we prove one of my favorite theorems.[1] It's a great example for recent Calculus BC students:

---

[1] I would like to say that I like this theorem because we can use it to prove Euler's Theorem in Complex Analysis. However, the true reason it brings a smile to my face is that Leon Simon once asked one of my friends, who was sleeping in class, to prove it. My friend was completely dumbfounded and Professor Simon retorted with his Australian accent, "Caught'cha napping!" Good times.

Example. *If a differentiable function f satisfies $f'(x) = 0$ for every x, then f is constant.*

Proof Summary:

- Suppose f is not constant.

- Apply Mean Value Theorem to two different points.

- Contradict $f'(x) = 0$.

Proof: Suppose f is not constant. Then f must have two different values at two points, say a and b, such that

$$f(a) \neq f(b)$$

Recall that the Mean Value Theorem states,

> *Given any open interval, we can find a point in this interval where the derivative of f is equal to the slope of the secant line.*

Formally, using interval (a, b) we can find $c \in (a, b)$ such that

$$f'(c) = \frac{f(b) - f(a)}{b - a}.$$

But $f'(c)$ is not zero since $f(b) \neq f(a)$! Thus, we have a contradiction, so $f(x)$ must be constant. □

In summary, Proof by Contradiction is one of the most powerful techniques in mathematics. In fact, there are some theorems that are **very hard** to prove if you do not allow proof by contradiction (the constructivists lurking in the math department can attest to this). Personally, I have no clue how to prove a number is irrational without using proof by contradiction!

Lastly, a tricky aspect of proof by contradiction is that

> *You may not know what contradiction you need to reach.*

In one example, we contradicted that a fraction was in reduced form. Whereas in another example, we contradicted $n = n$. The possible contradictions you can form are way out there: for example, in the proof that e is irrational, you create a number that is both an integer and not an integer! But *finding* what you need to contradict is one of the things that makes mathematics an **art**.

3.6   Proof Technique: If and Only If

Me: If you are good at Physics, then you are good at Calculus.
Dhruv: But I suck at Physics.
Me: But you can still be good at Calculus!

In Calculus BC, the whole concept of a limit was very informal. You knew it intuitively, but you never had to prove it formally. The same was true of the double arrow sign:

$$\iff$$

You knew it meant "equivalently," but you never talked about it formally. The most you probably did was fill in some truth table. But,

> Math Mantra: If you want to study mathematics, you have to be precise and
> everything you do must have meaning!

Let A and B be statements. Then,

$$A \Rightarrow B$$

means

Assuming A you can **derive** the outcome B.

Likewise,

$$A \Leftarrow B$$

means

Assuming B you can **derive** the outcome A.

For example, let A and B represent

$$A : \ x \ is \ a \ prime \ greater \ than \ 2$$
$$B : \ x \ is \ odd$$

Then $A \Rightarrow B$ means

If you assume x is a prime number greater than 2, then you can derive the outcome x is odd.

The double arrow notation

$$A \iff B$$

means A is equivalent to B in the sense that,

If you **assume** *A, then you can* **derive** *B*
and
if you **assume** *B, then you can* **derive** *A.*

So logically, statement A is really the same as statement B.

For example, consider the implication:

$$x = 1 \Rightarrow x^2 = 1$$

This is true. However, it is **false** to write

$$x = 1 \iff x^2 = 1$$

If we start at $x = 1$, then it is the case that $x^2 = 1$. But we cannot go backwards:[1] even if $x^2 = 1$, it is possible that $x = -1$.

However, the following is true:

$$x + y = 2 \iff x = 2 - y.$$

This is because you can perform an *invertible operation* to get from one equation to another.

Here are three proofs that use *if and only if*:

Example. *An integer n is even if and only if $n + 1$ is odd.*

Proof:

- $\Rightarrow$

  Let n be even. Then

  $$n = 2k$$

  for some integer k. But then

  $$n + 1 = 2k + 1$$

  which is the definition of odd.

- $\Leftarrow$

  Let $n + 1$ be odd. Then

  $$n + 1 = 2k + 1$$

  for some integer k. But then

  $$n = 2k$$

  which is again even by definition. $\qquad\qquad\qquad\qquad\qquad\qquad$ □

Notice we could have compressed both directions into a single proof. This is often possible when each step in the proof is invertible.

---

[1]Unless we are working with non-negative numbers.

Proof (Compressed):

$$
\begin{aligned}
n &= 2k && \text{for some integer } k \iff \\
n+1 &= 2k+1 && \text{for some integer } k
\end{aligned}
$$
$\square$

The next example requires a *fundamental* fact about the integers:

Fundamental Theorem of Arithmetic: *Every integer $n \geq 2$ has a unique prime factorization (up to reordering the multiplication):*

$$n = p_1^{\alpha_1} p_2^{\alpha_2} \ldots p_r^{\alpha_r}$$

for some distinct primes $p_1, p_2, \ldots, p_r$ and integer powers $\alpha_1, \alpha_2, \ldots, \alpha_r \geq 0$.

For example,

$$2250 = 2 \cdot 3^2 \cdot 5^3.$$

Example. *A positive integer is a perfect square if and only if every power in its prime factorization is even.*

Proof:

- $\Rightarrow$

  Let n be a perfect square. The $n = k^2$ by definition of a perfect square. But k has a prime factorization by the Fundamental Theorem of Arithmetic. So,

  $$k = p_1^{\alpha_1} p_2^{\alpha_2} \ldots p_r^{\alpha_r}$$

  for some distinct primes $p_1, p_2, \ldots, p_r$ and powers $\alpha_1, \alpha_2, \ldots, \alpha_r$. Thus,

  $$n = k^2 = (p_1^{\alpha_1} p_2^{\alpha_2} \ldots, p_r^{\alpha_r})^2 = p_1^{2\alpha_1} p_2^{2\alpha_2} \ldots p_r^{2\alpha_r}$$

  is *the* prime factorization of n and it contains only even powers.

- $\Leftarrow$

  Let n have only even powers in its prime factorization. Then,

  $$n = p_1^{2\alpha_1} p_2^{2\alpha_2} \ldots p_r^{2\alpha_r}$$

  for some distinct primes $p_1, p_2, \ldots, p_r$ and powers $\alpha_1, \alpha_2, \ldots, \alpha_r$. But then,

  $$n = p_1^{2\alpha_1} p_2^{2\alpha_2} \ldots p_r^{2\alpha_r} = (p_1^{\alpha_1} p_2^{\alpha_2} \ldots p_r^{\alpha_r})^2 = k^2$$

  for integer $k = p_1^{\alpha_1} p_2^{\alpha_2} \ldots p_r^{\alpha_r}$. Thus n is a perfect square. $\square$

Once again, the proof is compressible:

Proof (Compressed):

$$
\begin{aligned}
n &= k^2 &&\Longleftrightarrow \\
n &= (p_1^{\alpha_1} p_2^{\alpha_2} \dots p_r^{\alpha_r})^2 &&\Longleftrightarrow \\
n &= p_1^{2\alpha_1} p_2^{2\alpha_2} \dots p_r^{2\alpha_r}
\end{aligned}
$$

$\square$

We can also complete the following proofs:

Example. *An integer $n \geq 2$ is composite if and only if n is divisible by some prime $p < n$.*

Proof:

- $\Rightarrow$

  We already proved this via contradiction.[1]

- $\Leftarrow$

  Since n is divisible by some prime $p < n$, n is composite by definition. $\square$

Example. *Given a differentiable function f, $f'(x) = 0$ for every x if and only if $f(x)$ is constant.*

Proof:

- $\Rightarrow$

  We already proved this via contradiction.

- $\Leftarrow$

  The derivative of a constant function is 0. $\square$

Sometimes *if and only if* will be a godsend: it will give you two equivalent statements that are useful at different times (you will see this in the discussion of closed sets and limits). Or it could be used to reduce some unwieldy monstrosity of a definition to an easy equivalent statement.

3.7 Linear Independence

Now that we have discussed two of the most important proof techniques, let's return to the original question:

When are vectors in our spanning list redundant?

---

[1] In fact, we proved something stronger: if n is composite then n is divisible by some prime $p \leq \sqrt{n}$.

We make the following definitions:

Definition. *We say a set of vectors $\vec{v}_1, \vec{v}_2, \ldots, \vec{v}_n \in \mathbb{R}^n$ is **linearly dependent** if there exists coefficients $c_1, c_2, \ldots, c_n$ **not all zero** such that*

$$c_1\vec{v}_1 + c_2\vec{v}_2 + \ldots + c_n\vec{v}_n = \vec{0}.$$

*We also say a set is **linearly independent** if that set is not linearly dependent.*

Using this definition, the answer to our original question is:

the spanning list has no redundant vector if and only if the spanning list is linearly independent.

Generally,

Theorem. *A set is linearly independent if and only if no vector in the set can be written as a linear combination of other vectors in the set.*

Proof:

- $\Rightarrow$

  Let $\vec{v}_1, \vec{v}_2, \ldots, \vec{v}_n$ be a linearly independent set of vectors. Suppose, for an eventual contradiction, that some vector $\vec{v}_i$ in the set can be written as a linear combination of the other vectors. Then, by reordering,[1] we can assume

  $$\vec{v} = \vec{v}_1$$

  and

  $$\vec{v}_1 = c_2\vec{v}_2 + c_3\vec{v}_3 + \ldots + c_n\vec{v}_n$$

  for some (possibly zero) $c_1, c_2, \ldots, c_n \in \mathbb{R}$. By moving all terms to one side,

  $$(-1)\vec{v}_1 + c_2\vec{v}_2 + c_3\vec{v}_3 + \ldots + c_n\vec{v}_n = \vec{0}$$

  But this is a non-trivial linear combination of the zero vector since $c_1 = -1$, contradicting linear independence. In conclusion, no vector in the set can be written as a linear combination of other vectors.

- $\Leftarrow$

  Assume no vector in the set can be written as a linear combination of other vectors in the set. Suppose the set is not linearly independent. Then the set is, by definition, linearly dependent and there exist constants $c_1, c_2, \ldots, c_n$ not all zero, such that

  $$c_1\vec{v}_1 + c_2\vec{v}_2 + \ldots + c_n\vec{v}_n = \vec{0}$$

---

[1]The jargon is WLOG, "without loss of generality." We use this expression when we claim that we can solve the *general case* through the solution of a *specific* case.

We know that one of these coefficients is non-zero, so without loss of generality, we can reorder these vectors so that c_1 is *non-zero*. Then, isolating $\vec{v}_1$,

$$\vec{v}_1 = -\frac{c_2}{c_1}\vec{v}_2 - \frac{c_3}{c_1}\vec{v}_3 - \ldots - \frac{c_n}{c_1}\vec{v}_n$$

But we have written one vector as a linear combination of the remaining vectors, contradicting our initial assumption. □

3.8 On Keeping One's Word: Cauchy-Schwarz Equality

Now that we have the tools, we can answer our Cauchy-Schwarz Question from Lecture 2:

When is equality achieved in the Cauchy-Schwarz inequality?

The answer: equality in Cauchy-Schwarz holds **if and only if** two vectors are *parallel*:

Theorem.

$$|\vec{x} \cdot \vec{y}| = \|\vec{x}\| \, \|\vec{y}\|$$

if and only if $\vec{x} = t\vec{y}$ for some $t \in \mathbb{R}$.

Proof:

- $\Leftarrow$

  Let $\vec{x} = t\vec{y}$ for some t in $\mathbb{R}$. Then, we need to show

  $$|\underbrace{t\vec{y}}_{\vec{x}} \cdot \vec{y}| = \|\underbrace{t\vec{y}}_{\vec{x}}\| \, \|\vec{y}\|.$$

  Starting from the left, use absolute value properties to rewrite

  $$|t\vec{y} \cdot \vec{y}| = |t(\vec{y} \cdot \vec{y})| = |t| \, |\vec{y} \cdot \vec{y}|$$

  Converting to norms, this is

  $$|t| \, |\vec{y} \cdot \vec{y}| = |t| \, \|\vec{y}\|^2.$$

  Now, pull the $|t|$ inside one of the norms:

  $$|t| \, \|\vec{y}\|^2 = |t| \, \|\vec{y}\| \, \|\vec{y}\| = \|t\vec{y}\| \, \|\vec{y}\|.$$

- $\Rightarrow$

  To show two vectors are equal, we can exploit the zero property:

  $$\|\vec{x}\| = 0 \iff \vec{x} = \vec{0}.$$

  Particularly, if we can find a t such that

  $$\|\vec{x} - t\vec{y}\| = 0$$

  then

  $$\vec{x} = t\vec{y}$$

  But how do we find such a t?

> Math Mantra: We assume[1] for the moment that something exists to extract
> information on what it MUST look like and form a guess. Then, with our guess,
> we retry the proof.

Let's assume, for the moment, a t exists such that

$$\|\vec{x} - t\vec{y}\| = 0$$

Since it is easier to deal with squares, we square both sides to get

$$
\begin{aligned}
\|\vec{x} - t\vec{y}\|^2 &= (\vec{x} - t\vec{y}) \cdot (\vec{x} - t\vec{y}) \\
&= \vec{x} \cdot \vec{x} - 2t\vec{x} \cdot \vec{y} + t^2 \vec{y} \cdot \vec{y} \\
&= \underbrace{\|\vec{x}\|^2}_{c} + \underbrace{-2\vec{x} \cdot \vec{y}t}_{bt} + \underbrace{t^2 \|\vec{y}\|^2}_{at^2} = 0
\end{aligned}
$$

Therefore, *if* such a t were to exist, it would be given by the quadratic expression above. Applying the quadratic formula,

$$
t = \frac{2\vec{x} \cdot \vec{y} \pm \sqrt{(2\vec{x} \cdot \vec{y})^2 - 4\|\vec{x}\|^2 \|\vec{y}\|^2}}{2\|\vec{y}\|^2} = \frac{\vec{x} \cdot \vec{y} \pm \sqrt{(\vec{x} \cdot \vec{y})^2 - \|\vec{x}\|^2 \|\vec{y}\|^2}}{\|\vec{y}\|^2}
$$

But we assumed Cauchy-Schwarz equality holds; thus, we can replace the inner dot product with the product of norms:

$$
t = \frac{\vec{x} \cdot \vec{y} \pm \overbrace{\sqrt{(\|\vec{x}\|\,\|\vec{y}\|)^2 - \|\vec{x}\|^2 \|\vec{y}\|^2}}^{=0}}{\|\vec{y}\|^2} = \frac{\vec{x} \cdot \vec{y}}{\|\vec{y}\|^2}.
$$

Are we done? **No.** We only solved for what t **must look like.** Now, we must erase all our previous work and **define**

$$
t = \frac{\vec{x} \cdot \vec{y}}{\|\vec{y}\|^2}.
$$

We then show that our *constructed* t satisfies

$$\|\vec{x} - t\vec{y}\| = 0$$

Starting from the left, we square to get

$$\|\vec{x} - t\vec{y}\|^2 = \|\vec{x}\|^2 - 2t\vec{x} \cdot \vec{y} + t^2 \|\vec{y}\|^2$$

Substituting t, we have

$$
\|\vec{x}\|^2 - 2 \underbrace{\left(\frac{\vec{x} \cdot \vec{y}}{\|\vec{y}\|^2} \right)}_{t} \vec{x} \cdot \vec{y} + \underbrace{\left(\frac{\vec{x} \cdot \vec{y}}{\|\vec{y}\|^2} \right)^2}_{t^2} \|\vec{y}\|^2
$$

---

[1]This is a cheat that we will use again when we see determinants.

which is the same as

$$\|\vec{x}\|^2 - 2\left(\frac{(\vec{x}\cdot\vec{y})^2}{\|\vec{y}\|^2}\right) + \frac{(\vec{x}\cdot\vec{y})^2}{\|\vec{y}\|^2}.$$

But we assumed Cauchy-Schwarz equality holds; thus, this is equal to

$$\|\vec{x}\|^2 - 2\left(\frac{(\|\vec{x}\|\|\vec{y}\|)^2}{\|\vec{y}\|^2}\right) + \frac{(\|\vec{x}\|\|\vec{y}\|)^2}{\|\vec{y}\|^2}.$$

Expanding, we now have

$$\|\vec{x}\|^2 - 2\|\vec{x}\|^2 + \|\vec{x}\|^2 = 0$$

Since we have shown that

$$\|\vec{x} - t\vec{y}\|^2 = 0$$

we conclude

$$\|\vec{x} - t\vec{y}\| = 0$$

and thus

$$\vec{x} = t\vec{y}.$$

Are we done now? No! Just like in the proof of Cauchy-Schwarz inequality,

There is a slight fly in the ointment!

Again, our definition of t requires $\|\vec{y}\| \neq \vec{0}$ which need not be the case. However, the proof is trivial in the case $\vec{y} = \vec{0}$ since

$$0\vec{x} = \vec{y}.$$

Now we're done! □

Let's use this result on our fun Cauchy-Schwarz examples in Lecture 2!

Example. *For any angle θ,*

$$|\cos^2\theta - \sin^2\theta| \leq 1.$$

Equality holds if and only if for some constant c

$$\begin{bmatrix} \cos\theta \\ \sin\theta \end{bmatrix} = \begin{bmatrix} c\cos\theta \\ c(-\sin\theta) \end{bmatrix}.$$

Equating components,

$$\begin{aligned} \cos\theta &= c\cos\theta \\ \sin\theta &= -c\sin\theta \end{aligned}$$

Equivalently,

$$\begin{aligned} (1-c)\cos\theta &= 0 \\ (1+c)\sin\theta &= 0 \end{aligned}$$

The first equation holds if and only if

$$c = 1 \quad \text{or} \quad \theta = \frac{\pi}{2} + \pi n \text{ for any integer } n$$

whereas the second equation requires

$$c = -1 \quad \text{or} \quad \theta = \pi n \text{ for any integer } n.$$

In the case $c = 1$, we can plug c into the second equation to get

$$2\sin\theta = 0$$

which holds if and only if

$$\theta = \pi n \text{ for any integer } n.$$

Likewise, plugging the case $c = -1$ into the first equation yields

$$2\cos\theta = 0$$

which is true if and only if

$$\theta = \frac{\pi}{2} + \pi n \text{ for any integer } n.$$

Therefore, *both* equations hold if and only if

$$\theta = \pi n \text{ for any integer } n \quad \text{or} \quad \theta = \frac{\pi}{2} + \pi n \text{ for any integer } n$$

In conclusion, $\left| \cos^2\theta - \sin^2\theta \right| = 1$ if and only if

$$\theta = \frac{\pi n}{2} \text{ for any integer } n.$$

Example. *For any $a, b, c > 0$, we have*

$$(a + b + c)\left(\frac{1}{a} + \frac{1}{b} + \frac{1}{c}\right) \geq 9$$

Equality holds if and only if, for some constant t,

$$\begin{bmatrix} \sqrt{a} \\ \sqrt{b} \\ \sqrt{c} \end{bmatrix} = \begin{bmatrix} \dfrac{t}{\sqrt{a}} \\ \dfrac{t}{\sqrt{b}} \\ \dfrac{t}{\sqrt{c}} \end{bmatrix}$$

Equating components,

$$\sqrt{a} = \frac{t}{\sqrt{a}}$$

$$\sqrt{b} = \frac{t}{\sqrt{b}}$$

$$\sqrt{c} = \frac{t}{\sqrt{c}}.$$

Isolating t in each equation, we see that equality holds if and only if

$$a = b = c.$$

Example. *Let $a_1, a_2, \ldots, a_n \in \mathbb{R}$, then*

$$\left(\frac{1}{n} \sum_{i=1}^{n} a_i \right)^2 \leq \frac{1}{n} \sum_{i=1}^{n} a_i^2.$$

Equality holds if and only if, for some constant c,

$$\begin{bmatrix} a_1 \\ a_2 \\ \vdots \\ a_n \end{bmatrix} = \begin{bmatrix} \dfrac{c}{n} \\ \dfrac{c}{n} \\ \vdots \\ \dfrac{c}{n} \end{bmatrix}.$$

But this is equivalently

$$a_1 = a_2 = \ldots = a_n.$$

Example. *Let $a_1, a_2, \ldots, a_n$ be positive. Then,*

$$a_1 + a_2 + \ldots + a_n \leq \frac{a_1^2}{a_2} + \frac{a_2^2}{a_3} + \ldots + \frac{a_{n-1}^2}{a_n} + \frac{a_n^2}{a_1}$$

Equality holds if and only if for some constant c

$$\begin{bmatrix} \sqrt{a_2} \\ \sqrt{a_3} \\ \vdots \\ \sqrt{a_n} \\ \sqrt{a_1} \end{bmatrix} = \begin{bmatrix} \dfrac{ca_1}{\sqrt{a_2}} \\ \dfrac{ca_2}{\sqrt{a_3}} \\ \vdots \\ \dfrac{ca_{n-1}}{\sqrt{a_n}} \\ \dfrac{ca_n}{\sqrt{a_1}} \end{bmatrix}.$$

This tells us

$$\begin{aligned} a_2 &= ca_1 \\ a_3 &= ca_2 \\ &\vdots \\ a_n &= ca_{n-1} \\ a_1 &= ca_n. \end{aligned}$$

Substituting each equality into the next yields

$$a_1 = c^n a_1.$$

Equivalently,

$$(1 - c^n)a_1 = 0.$$

Thus,

$$a_1 = 0 \text{ or } c = \pm 1.$$

But $a_1 \neq 0$ by definition. Moreover, $c \neq -1$ since $a_2 = ca_1$ and a_2, a_1 are both positive. Therefore, equality only holds when and only when $c = 1$. Equivalently, when $c = 1$,

$$a_1 = a_2 = \ldots = a_n.$$

New Notation

Symbol	Reading	Example	Example Translation
$A \cap B$	The intersection of sets A and B.	$\mathbb{Q} \cap \mathbb{R} = \mathbb{Q}$	The intersection of the set of reals and the set of rationals is the set of rationals.
$\text{span}\{\vec{v}_1, \vec{v}_2, \ldots, \vec{v}_n\}$	The span of the set of vectors $\vec{v}_1, \vec{v}_2, \ldots, \vec{v}_n$	$\text{span}\{\vec{v}_1\} \subseteq \mathbb{R}^2$	The span of $\vec{v}_1$ is contained in $\mathbb{R}^2$.
$A \Rightarrow B$	A implies B	$x = 1 \Rightarrow x^2 = 1$	If a number is equal to one, then its square is 1.
$A \Leftarrow B$	B implies A	$x = 0 \text{ or } y = 0 \Leftarrow xy = 0$	If the product of two numbers is zero, then at least one of them is zero.
$A \Longleftrightarrow B$	A if and only if B	$x = y + 2 \Longleftrightarrow x - 2 = y$	$x = y + 2$ if and only if $x - 2 = y$

Lecture 4

Under-determined Potential

Think around the statement (the Under-determined Systems Lemma). Try to uncover its meaning and how one could perhaps approach a proof of it. Look at special cases to see if they shed light on the statement. Maybe the statement is false!

-Leon Simon

Goals: Today, we focus on systems of homogeneous linear equations. In order to simplify them without changing the solutions, we apply the first step of Gaussian Elimination. This will be the key step in proving the Under-determined Systems Lemma. We will also need to apply two new proof methods: the intuitive method of *proof by cases*, and the often mistaught *proof by induction*. We then use the Under-determined Systems Lemma to prove the incredibly important Linear Dependence Lemma.

4.1 System of Equations

In Algebra II, you learned how to solve a system of equations like

$$
\begin{aligned}
3x &+ 2y &+ z &= 1 \\
2x &+ y &+ 4z &= 3 \\
5x &+ 2y &- z &= 0
\end{aligned}
$$

Of course, we can generalize to a system with m equations with n unknowns:

$$
\begin{aligned}
a_{11}x_1 &+ a_{12}x_2 &+ a_{13}x_3 &+ \ldots &+ a_{1n}x_n &= b_1 \\
a_{21}x_1 &+ a_{22}x_2 &+ a_{23}x_3 &+ \ldots &+ a_{2n}x_n &= b_2 \\
&\vdots & \vdots & \vdots & \vdots & \vdots \\
a_{m1}x_1 &+ a_{m2}x_2 &+ a_{m3}x_3 &+ \ldots &+ a_{mn}x_n &= b_m
\end{aligned}
$$

Here, a_{ij} is the coefficient[1] of the x_j term in the i-th equation and b_i is the right-hand side of the i-th equation.

---

[1]Sometimes I wish the notation would be $a_i{}^j$. Especially since, for example, $a_{(m+1)2}$ can be misconstrued as $a_{(2m+2)}$. Nevertheless, we reserve such superscript notation for exponents and sequences.

In this lecture we are going to consider only systems with each $b_i = 0$. We call such system *homogeneous*:

$$
\begin{array}{ccccccccccc}
a_{11}x_1 & + & a_{12}x_2 & + & a_{13}x_3 & + & \ldots & + & a_{1n}x_n & = & 0 \\
a_{21}x_1 & + & a_{22}x_2 & + & a_{23}x_3 & + & \ldots & + & a_{2n}x_n & = & 0 \\
\vdots & & \vdots & & \vdots & & \vdots & & \vdots & & \vdots \\
a_{m1}x_1 & + & a_{m2}x_2 & + & a_{m3}x_3 & + & \ldots & + & a_{mn}x_n & = & 0
\end{array}
$$

Why are we considering only homogeneous systems instead of general (possibly inhomogeneous) systems?

- **We are always guaranteed a solution.**

  Namely, we always have the trivial solution

  $$x_1 = x_2 = \ldots = x_n = 0$$

  Thus, we don't have to worry about having no solutions which as you recall, arises from inconsistent equations like

  $$
  \begin{array}{ccccc}
  x_1 & + & 2x_2 & = & 1 \\
  x_1 & + & 2x_2 & = & 0
  \end{array}
  $$

  If such x_1, x_2 did exist, then $1 = 0$, which is a big no-no.

- **The homogeneous solution is needed to construct the solution of an inhomogeneous system!**

  Without going into too much detail, the reason is analogous to linear functions. Consider a linear function f that satisfies

  $$f(y) = b$$

  Let y' be a solution to the homogeneous analogue

  $$f(y') = 0.$$

  Then,

  $$f(y + y') = f(y) + f(y') = 0 + b = b.$$

  This means the sum of a solution of the inhomogeneous system and of the homogeneous system is still a solution of the inhomogeneous system!

- **We will only need homogeneous systems to prove the almighty Linear Dependence Lemma.**

The first fact we will prove about homogeneous systems is the Under-determined Systems Lemma. But to prove this fact, we first need to talk about the Gaussian Elimination process. Here, we only focus on the first step, but in **Lecture 12** we will use the full force of Gaussian Elimination to compute reduced row echelon forms.

4.2 Gaussian Elimination: Step One

The first step of Gaussian Elimination is used to help determine the solution to the first unknown, x_1. This is *exactly* what you did back in those matrix days, when simplifying

$$\left[\begin{array}{ccc|c} 0 & 2 & 0 & 1 \\ 2 & 1 & 0 & 2 \\ 3 & 1 & 1 & 3 \end{array}\right]$$

Except we are going to be a lot more formal about it. Particularly, we are going to reserve matrix notation for **Lecture 7**. So pretend, for the moment, that matrices don't exist and just phrase everything in terms of systems of equations.

Consider the system

$$
\begin{array}{ccccccccc}
a_{11}x_1 & + & a_{12}x_2 & + & a_{13}x_3 & + & \ldots & + & a_{1n}x_n & = & 0 \\
a_{21}x_1 & + & a_{22}x_2 & + & a_{23}x_3 & + & \ldots & + & a_{2n}x_n & = & 0 \\
\vdots & & \vdots & & \vdots & & \vdots & & \vdots & & \\
a_{m1}x_1 & + & a_{m2}x_2 & + & a_{m3}x_3 & + & \ldots & + & a_{mn}x_n & = & 0
\end{array}
$$

The **first step of Gaussian Elimination** performs one of the two actions:

- CASE 1: If the coefficient a_{i1} of x_1 in each equation is 0 (the first "column" contains all zeros),

$$
\begin{array}{ccccccccc}
0x_1 & + & a_{12}x_2 & + & a_{13}x_3 & + & \ldots & + & a_{1n}x_n & = & 0 \\
0x_1 & + & a_{22}x_2 & + & a_{23}x_3 & + & \ldots & + & a_{2n}x_n & = & 0 \\
\vdots & & \vdots & & \vdots & & \vdots & & \vdots & & \\
0x_1 & + & a_{m2}x_2 & + & a_{m3}x_3 & + & \ldots & + & a_{mn}x_n & = & 0
\end{array}
$$

  you are done.

- CASE 2: Otherwise, in some equation k, the coefficient $a_{k1} \neq 0$. Swap the first and k-th equations:

$$
\begin{array}{ccccccccc}
a_{k1}x_1 & + & a_{k2}x_2 & + & a_{k3}x_3 & + & \ldots & + & a_{kn}x_n & = & 0 \\
a_{21}x_1 & + & a_{22}x_2 & + & a_{23}x_3 & + & \ldots & + & a_{2n}x_n & = & 0 \\
\vdots & & \vdots & & \vdots & & \vdots & & \vdots & & \\
a_{11}x_1 & + & a_{12}x_2 & + & a_{13}x_3 & + & \ldots & + & a_{1n}x_n & = & 0 \\
\\
\vdots & & \vdots & & \vdots & & \vdots & & \vdots & & \\
a_{m1}x_1 & + & a_{m2}x_2 & + & a_{m3}x_3 & + & \ldots & + & a_{mn}x_n & = & 0
\end{array}
$$

  SWAP

  Then divide the first equation by a_{k1} to get

$$
\mathbf{1}x_1 \; + \; \frac{a_{k2}}{a_{k1}}x_2 \; + \; \frac{a_{k3}}{a_{k1}}x_3 \; + \; \ldots \; + \; \frac{a_{kn}}{a_{k1}}x_n \; = \; 0.
$$

- Subtract the proper multiples of the first equation from each of the remaining equations, so that each coefficient of x_i is zero:

$$
\begin{aligned}
1x_1 &+ a'_{12}x_2 &+ a'_{13}x_3 &+ \ldots &+ a'_{1n}x_n &= 0 \\
0x_1 &+ a'_{22}x_2 &+ a'_{23}x_3 &+ \ldots &+ a'_{2n}x_n &= 0 \\
&\ \ \vdots &\vdots &\vdots &\vdots &\vdots \\
0x_1 &+ a'_{m2}x_2 &+ a'_{m3}x_3 &+ \ldots &+ a'_{mn}x_n &= 0
\end{aligned}
$$

Note the new a'_{ij}: we simply relabeled coefficients because we do not want to be distracted by fractions and subtractions. We **only care** about the *structure* of the **first column**.

Example. *Compute the first step of Gaussian Elimination for*

$$
\begin{aligned}
0x_1 &+ 2x_2 &+ 1x_3 &= 0 \\
2x_1 &+ 2x_2 &+ 4x_3 &= 0 \\
6x_1 &+ 2x_2 &- 1x_3 &= 0
\end{aligned}
$$

We switch the second row with the first, yielding

$$
\begin{aligned}
2x_1 &+ 2x_2 &+ 4x_3 &= 0 \\
0x_1 &+ 2x_2 &+ 1x_3 &= 0 \\
6x_1 &+ 2x_2 &- 1x_3 &= 0
\end{aligned}
$$

Then, divide the first equation by the coefficient of x_1:

$$
\begin{aligned}
1x_1 &+ 1x_2 &+ 2x_3 &= 0 \\
0x_1 &+ 2x_2 &+ 1x_3 &= 0 \\
6x_1 &+ 2x_2 &- 1x_3 &= 0
\end{aligned}
$$

Finally, subtract some multiple of the first equation from each of the remaining equations, so that their x_1 coefficients are 0:

$$
\begin{aligned}
1x_1 &+ 1x_2 &+ 2x_3 &= 0 \\
0x_1 &+ 2x_2 &+ 1x_3 &= 0 \\
0x_1 &- 4x_2 &- 13x_3 &= 0
\end{aligned}
$$

In summary, the **key observation** is that after the first step of Gaussian Elimination, our system is transformed to one of two possible forms:

$$
\begin{array}{|llll|}
\hline
0x_1 + a'_{12}x_2 + \ldots + a'_{1n}x_n = 0 \\
0x_1 + a'_{23}x_3 + \ldots + a'_{2n}x_n = 0 \\
\ \ \vdots \quad\ \ \vdots \quad\ \ \vdots \quad\ \ \vdots \\
0x_1 + a'_{m2}x_2 + \ldots + a'_{mn}x_n = 0 \\
\hline
\end{array}
\qquad
\begin{array}{|llll|}
\hline
1x_1 + a'_{12}x_2 + \ldots + a'_{1n}x_n = 0 \\
0x_1 + a'_{23}x_3 + \ldots + a'_{2n}x_n = 0 \\
\ \ \vdots \quad\ \ \vdots \quad\ \ \vdots \quad\ \ \vdots \\
0x_1 + a'_{m2}x_2 + \ldots + a'_{mn}x_n = 0 \\
\hline
\end{array}
$$

By the way, you should ask yourself:

Does Gaussian Elimination preserve the solution space? Does it create or destroy solutions?

Intuitively, it is clear that the solutions are unchanged, but we should provide a proof. However, such a proof requires that we prove two sets are equal, a technique I shall delay until more urgent need arises.

The proof techniques that **need** immediate attention, however, are the intuitive *proof by cases* and the less intuitive *induction*. If you want to understand the Under-determined Systems Lemma (as well as many proofs in this course), you need to learn how to apply both techniques.

4.3 Proof Technique: Proof by Cases

All Roads Lead to Rome.

-Proverb

Consider the story of poor Jose, who grew up in rural Portugal. Each day, his mother would ask him and his seven siblings,

Would you like chicken, fish, or beef for dinner?

Regardless of what the children chose, the answer from the mother would always be the same:

We are having fish.

This is all *Proof by Cases* really is: every possible choice leads to the same result. Precisely,

If every possibility implies the same outcome, then that outcome MUST be true.

Visually you can think of a crossroads; no matter which path you take, you end up at the same place.

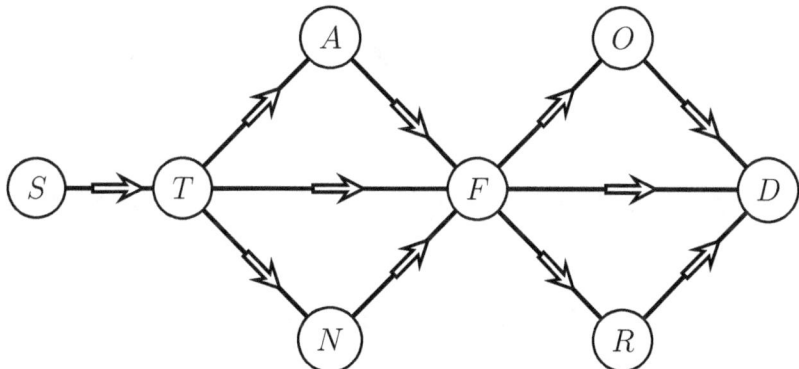

We give the prototypical example of proof by cases (and the bane of the mathematical constructivists).

Example. *There exist irrational numbers a, b such that a^b is rational.*

Proof: We know that either

$$\sqrt{2}^{\sqrt{2}} \text{ is rational} \quad or \quad \sqrt{2}^{\sqrt{2}} \text{ is irrational}$$

- CASE 1: $\sqrt{2}^{\sqrt{2}}$ is rational.

  In this case, we are already done since we know $\sqrt{2}$ is irrational. Just choose

  $$
  \begin{aligned}
  a &= \sqrt{2} \\
  b &= \sqrt{2}
  \end{aligned}
  $$

- CASE 2: $\sqrt{2}^{\sqrt{2}}$ is irrational.

  In this case, we choose irrationals

  $$
  \begin{aligned}
  a &= \sqrt{2}^{\sqrt{2}} \\
  b &= \sqrt{2}
  \end{aligned}
  $$

  Then, by exponent properties,

  $$
  \left(\sqrt{2}^{\sqrt{2}} \right)^{\sqrt{2}} = \sqrt{2}^{\sqrt{2} \cdot \sqrt{2}} = \sqrt{2}^2 = 2
  $$

  which is rational.

  Thus by cases, there exist irrational numbers a, b such that a^b is rational. $\square$

Did we actually find numbers a, b such that a^b is rational? Nope. **But we did prove such numbers exist!** This is one of those fun math phenomena: we can prove the existence of an object **without actually constructing it.** Love it or hate it, you must admit it is pretty cool.

Another fun example is one from 8th grade Geometry: recall (a, b, c) is a Pythagorean triple if

$$
a^2 + b^2 = c^2
$$

and a, b, c are all positive integers. We can prove that any integer $n \geq 3$ occurs in some Pythagorean triple.

For example, consider the number 17. We can find a Pythagorean triple containing it, namely $(144, 17, 145)$ since

$$
144^2 + 17^2 = 145^2.
$$

Example. *Any integer $n \geq 3$ occurs in some Pythagorean triple.*

Proof: We know that either

$$
n \text{ is odd} \quad or \quad n \text{ is even.}
$$

- CASE 1: n is odd.

  In this case, n^2 is odd. This means
  $$n^2 = 2q + 1$$
  for some positive integer q. Using this q
  $$q^2 + \underbrace{(2q + 1)}_{n^2} = (q + 1)^2.$$

  Therefore,
  $$(q, n, q + 1)$$
  is a Pythagorean triple containing n.

- CASE 2: n is even.

  In this case, $n^2 = 4q$ for some positive integer q. Using this q,
  $$(q - 1)^2 + \underbrace{(4q)}_{n^2} = (q + 1)^2$$

  Thus,
  $$(q - 1, n, q + 1)$$
  is a Pythagorean triple containing n. $\qquad\square$

Here is a fun example from a math competition:

> *On the planet Ianoia, the currency is in \$5 and \$11 bills. What is the largest dollar amount that cannot be created using only \$5 and \$11 bills?*

We can show that 39 cannot be created. Suppose it can. Then for some integers $x, y \geq 0$:

$$39 = 5x + 11y \implies 39 - 5x = 11y$$

But when we subtract multiples of 5 from 39, we never get a non-negative multiple of 11:

$$
\begin{aligned}
39 - 0 \cdot 5 &= 39 \\
39 - 1 \cdot 5 &= 34 \\
39 - 2 \cdot 5 &= 29 \\
39 - 3 \cdot 5 &= 24 \\
39 - 4 \cdot 5 &= 19 \\
39 - 5 \cdot 5 &= 14 \\
39 - 6 \cdot 5 &= 9 \\
39 - 7 \cdot 5 &= 4
\end{aligned}
$$

If we were in the contest, we might just guess that 39 is the largest amount that cannot be created with \$5 and \$11 bills. As aspiring mathematicians, however, we need to formally show that any dollar amount greater than 39 *can* be made with \$5 and \$11 bills:

Example. *Any integer $n \geq 40$ can be written as*

$$n = 5x + 11y$$

for some non-negative x, y.

Proof: We generalize the idea that n is odd or n is even. For some non-negative integer q, n must have one of the following forms:

$$
\begin{aligned}
n &= 5q \\
n &= 5q + 1 \\
n &= 5q + 2 \\
n &= 5q + 3 \\
n &= 5q + 4
\end{aligned}
$$

Why is this true? Simple: when you divide by 5, you either have remainder 0,1,2,3, or 4. Also notice, since $n \geq 40$, we have $q \geq 8$.

- CASE 1: $n = 5q$.

  In this case, we choose

  $$
  \begin{aligned}
  x &= q \\
  y &= 0.
  \end{aligned}
  $$

  where[1] $x \geq 8$.

- CASE 2: $n = 5q + 1$.

  Then,

  $$n = 5q + 1 = 5(q - 2) + 10 + 1 = 5(q - 2) + 11$$

  Choose

  $$
  \begin{aligned}
  x &= q - 2 \\
  y &= 1.
  \end{aligned}
  $$

  where $x \geq 8 - 2 = 6$.

- CASE 3: $n = 5q + 2$.

  Then,

  $$n = 5q + 2 = 5(q - 4) + 20 + 2 = 5(q - 4) + (11)2$$

  Choose

  $$
  \begin{aligned}
  x &= q - 4 \\
  y &= 2.
  \end{aligned}
  $$

  where $x \geq 8 - 4 = 4$.

---

[1] In each case, we need to check that x is non-negative. Indeed, this is where the argument would fail for some $n < 40$.

- CASE 4: $n = 5q + 3$.

  Then,
  $$n = 5q + 3 = 5(q - 6) + 30 + 3 = 5(q - 6) + 3(11)$$
  Choose
  $$\begin{aligned} x &= q - 6 \\ y &= 3 \end{aligned}$$
  where $x \geq 8 - 6 = 2$.

- CASE 5: $n = 5q + 4$.

  Then,
  $$n = 5q + 4 = 5(q - 8) + 40 + 4 = 5(q - 8) + 4(11)$$
  Choose
  $$\begin{aligned} x &= q - 8 \\ y &= 4 \end{aligned}$$
  where $x \geq 8 - 8 = 0$.

Thus, in each case, we can find non-negative x, y such that
$$n = 5x + 11y$$
$\square$

In general,

```
Math Mantra:  If you are stuck on a proof, try breaking it up into cases.  The
additional information you gain from each individual case can help you solve the
                     problem one step at a time.
```

4.4   Proof Technique: Induction

I am very powerful. Yet I am the least of all the guards. From hall to hall, door after door, each guard is more powerful than the last...

-Franz Kafka, *The Trial*

In all my years of teaching, I have found that students struggle most with this technique. But if you are going to work with n dimensional things or anything built by a recursive process, you **must** know induction. Before I tell you the details, let us talk about induction *algebraically*.

Consider the following game:

<u>THE GAME OF MODUS PONENS</u>

<u>OBJECTS:</u>

The only objects we will use are letters or two letters joined by the symbol $\Rightarrow$. For example,

$$X$$
$$Y$$
$$Z$$
$$X \Rightarrow Z$$
$$D \Rightarrow E$$
$$C \Rightarrow C$$

are all objects in our game. Usually, we will number letters with subscripts:

$$A_1$$
$$A_2$$
$$\vdots$$
$$A_n$$

<u>THE RULE:</u>

Suppose, we are given two objects
$$A \Rightarrow B$$

and
$$A.$$

Then we are allowed to create the brand new object

$$B$$

As an example of how to play this game, suppose you are given

$$\begin{aligned} A &\Rightarrow B \\ B &\Rightarrow C \\ B &\Rightarrow D \\ A & \end{aligned}$$

Using these objects, you can build the additional objects, B, C, D:

First, apply the rule to

$$\begin{aligned} A &\Rightarrow B \\ A & \end{aligned}$$

to get B. Now we apply the rule with B and the objects

$$B \Rightarrow C$$
$$B \Rightarrow D$$

respectively to get C and D.

Now, consider a game where you are given the object

$$A_1$$

and

$$A_n \Rightarrow A_{n+1}$$

for *any* natural number n. Note that the preceding line encodes infinitely many objects; namely:

$$A_1 \Rightarrow A_2$$
$$A_2 \Rightarrow A_3$$
$$A_3 \Rightarrow A_4$$
$$A_4 \Rightarrow A_5$$
$$\vdots$$

Using our rule, what objects can you build?

First, you can use

$$A_1 \Rightarrow A_2$$
$$A_1$$

to build A_2. Then, using

$$A_2 \Rightarrow A_3$$
$$A_2$$

you can build A_3. In fact, you can build A_n for *any* natural number n.

This is what induction *really* is: the **objects are true statements** like

$I:$ 0 is neither positive nor negative.
$A:$ John went to the movies.
$N:$ There is no smallest positive real number.

In particular, the objects containing $\Rightarrow$ are just true conditionals.

For example, the following are true:

$S \Rightarrow T:$ **If** $\overbrace{\text{you scored a 2400 on the SAT}}^{S}$, **then** $\overbrace{\text{you scored an 800 on the Math Section}}^{T}$.

$A \Rightarrow B:$ **If** $\overbrace{|\vec{x} \cdot \vec{y}| = \|\vec{x}\| \, \|\vec{y}\|}^{A}$, **then** $\overbrace{\vec{x} = t\vec{y}}^{B}$ for some t.

$C \Rightarrow D:$ **If** $\overbrace{\vec{v}_1, \vec{v}_2, \ldots, \vec{v}_n \text{ is a linearly independent set,}}^{C}$

then $\overbrace{\text{no } \vec{v}_i \text{ in this set can be written as a linear combination of the other vectors}}^{D}$.

Our "rule" is just the law of **Modus Ponens**:

MODUS PONENS: *If A is true and A implies B, then we can conclude B is true.*

For example, if we have

$$A \Rightarrow B \quad : \quad \text{If } \overbrace{n \text{ is prime}}^{A}, \text{ then } \overbrace{n \text{ is not } 8}^{B}.$$
$$A \qquad\quad : \quad n \text{ is prime.}$$

Then we conclude B:

$$B : \quad n \text{ is not } 8.$$

An object A_n is simply a true statement that is a **function** of n. And if we can prove **the base case**

$$A_1$$

and **the inductive step**

$$A_n \quad \Rightarrow \quad A_{n+1} \; ,$$

for any non-negative integer n, then, as in our game, we can conclude A_n is true for any natural number n.

For example, suppose we know

$$A_1 \quad : \quad 1 \text{ is greater than } 0$$

and for all natural numbers n,

$$A_n \Rightarrow A_{n+1} \quad : \quad \text{if } n \text{ is greater than } 0, \text{ then } n+1 \text{ is greater than } 0.$$

Then we equivalently know

$$A_1 \qquad\quad : \quad 1 \text{ is greater than } 0$$
$$A_1 \Rightarrow A_2 \quad : \quad \text{if } 1 \text{ is greater than } 0, \text{ then } 2 \text{ is greater than } 0.$$
$$A_2 \Rightarrow A_3 \quad : \quad \text{if } 2 \text{ is greater than } 0, \text{ then } 3 \text{ is greater than } 0.$$
$$\vdots \qquad\qquad\qquad\qquad \vdots$$

Using Modus Ponens, we conclude

$$A_1 \quad : \quad 1 \text{ is greater than } 0$$
$$A_2 \quad : \quad 2 \text{ is greater than } 0$$
$$A_3 \quad : \quad 3 \text{ is greater than } 0$$
$$\vdots \qquad\qquad \vdots$$

i.e. for all integer n

$$A_n \quad : \quad n \text{ is greater than } 0$$

Here are some examples of induction. However, I reserve the prototypical example of Gauss' lemma (and the warped high school monstrosity version of it), for the next section.

We call an object a *binary string* if it is a finite sequence of 1's and 0's. For example

$$0100010$$

is a binary string. We can also form new strings by *concatenating* them i.e placing one string after the other. Thus,

$$0\underline{1000}10\underline{0}10\underline{0}$$

is a string formed from concatenating the five strings

$$0, 100, 010, 010, 0$$

Using induction, we can formally prove the following intuitive result:

Example. *A string formed from concatenations of* 0, 100, *and* 010 *will have more* 0*'s than* 1*'s.*

Proof: We do induction on the number of string pieces that are concatenated. Formally, we want to prove, for every n, the property:

P_n : *A string formed from concatenating n pieces of* 0, 100, *and* 010 *has more* 0*'s than* 1*'s.*

- BASE CASE, $n = 1$.

  Suppose we have a final string formed from only one piece. Then it is either

  $$0, 100, 010$$

  which each have more 0's than 1's. Thus, P_1 is true.

- INDUCTIVE STEP

  Let n be an arbitrary natural number. To prove[1] $P_n \Rightarrow P_{n+1}$, we need to *assume P_n* and show that we can conclude

  P_{n+1} : *A string formed from concatenating $n + 1$ pieces of* 0, 100, *and* 010 *has more* 0*'s than* 1*'s.*

  Let S' be some arbitrary string built from $n + 1$ pieces. By definition of how we construct new strings, it must be formed from an n-piece string S, along with one of the strings

  $$0, 100, 010.$$

  Then, by our *inductive hypothesis* (our assumption of P_n), we know S has x zeros and y ones where $x > y$. So S' is either

---

[1] Indeed, to prove a conditional is true, you assume the *if* statement and show the *then* statement holds. In fact, you've already done this for several proofs!

String	Number of 0's	Number of 1's
$S0$	$x + 1$	y
$S100$	$x + 2$	$y + 1$
$S010$	$x + 2$	$y + 1$

In each case, the number of 0's is greater than the number of 1's since $x > y$, so P_{n+1} is true. Since n was *arbitrary*, we proved

$$P_n \Rightarrow P_{n+1}$$

for all natural numbers n. We can thus conclude by induction, that for $n \geq 1$, P_n holds i.e. a string formed from concatenating n pieces of 0, 100, and 010 has more zeros than ones. $\square$

Too easy? Here is a less obvious example:

Example. $8^n - 3^n$ *is divisible by 5 for every non-negative integer n.*

Proof: We do induction on the natural number n where

$$P_n : \quad 8^n - 3^n \text{ is divisible by 5}$$

- BASE CASE, $n = 0$

  P_0 is obviously true since

  $$8^0 - 3^0 = 0$$

  is divisible by 5.

- INDUCTIVE STEP

  Let n be an arbitrary non-negative integer. Assume

  $$P_n : \quad 8^n - 3^n \text{ is divisible by 5.}$$

We want to show

$$P_{n+1} : \quad 8^{n+1} - 3^{n+1} \text{ is divisible by 5}$$

is true. By algebraic shenanigans, we can rewrite

$$
\begin{aligned}
8^{n+1} - 3^{n+1} &= 8 \cdot 8^n - 3 \cdot 3^n \\
&= (5 + 3) \cdot 8^n - 3 \cdot 3^n \\
&= 5 \cdot 8^n + 3 \cdot 8^n - 3 \cdot 3^n \\
&= 5 \cdot 8^n + 3 \cdot (8^n - 3^n)
\end{aligned}
$$

By the inductive hypothesis,

$$8^n - 3^n = 5q$$

for some non-negative integer q. Thus,

$$
\begin{aligned}
8^{n+1} - 3^{n+1} &= 5 \cdot 8^n + 3 \cdot 5q \\
&= 5(8^n + 3q)
\end{aligned}
$$

meaning $8^{n+1} - 3^{n+1}$ is divisible by 5. Therefore, P_{n+1} is true. We can thus conclude by induction that for every non-negative integer n,

$$
8^n - 3^n \text{ is divisible by 5} \qquad \square
$$

Here is an induction example involving inequalities: consider the sequence defined by

$$
\begin{aligned}
S_1 &= 1 \\
S_k &= \sqrt{1 + S_{k-1}} \text{ for } k \geq 2
\end{aligned}
$$

The first few terms are

$$
\begin{aligned}
S_1 &= 1 \\
S_2 &= \sqrt{1 + \sqrt{1}} \\
S_3 &= \sqrt{1 + \sqrt{1 + \sqrt{1}}} \\
S_4 &= \sqrt{1 + \sqrt{1 + \sqrt{1 + \sqrt{1}}}}
\end{aligned}
$$

It turns out that this sequence converges to the golden ratio! We will prove this result in two weeks. But one of the key steps in that proof is the following:

Example. *$S_n < 2$ for all positive integer n, where we define S_k by*

$$
\begin{aligned}
S_1 &= 1 \\
S_k &= \sqrt{1 + S_{k-1}} \text{ for } k \geq 2
\end{aligned}
$$

Informally, we are showing

$$
\sqrt{1 + \sqrt{1 + \sqrt{1 + \sqrt{\ldots \sqrt{1 + \sqrt{1}}}}}} < 2
$$

no matter how many finite iterations the "$\ldots$" represent.

Proof: We do induction on the term number (the subscript of S). That is, for each $n \geq 1$, we take property P_n to be

$$
P_n : \quad S_n < 2
$$

We will show P_n is true for all $n \geq 1$.

- BASE CASE, $n = 1$:

  P_1 is obviously true since $1 < 2$.

- INDUCTIVE STEP

  Let $n \geq 1$ be arbitrary. Assume that property P_n is true. We want to show

  $$P_{n+1}: \quad S_{n+1} < 2$$

  is true. But we know

  $$\begin{aligned} S_{n+1} &= \sqrt{1 + S_n} && \text{(by Recursive Definition)} \\ &< \sqrt{1+2} = \sqrt{3} && \text{(by Inductive Hypothesis)} \\ &< 2 \end{aligned}$$

  Thus P_{n+1} is true. By induction, we conclude for every natural number n,

  $$P_n: \quad S_n < 2 \qquad\qquad \square$$

Finally, we provide an alternate proof of the Cauchy-Schwarz inequality using induction. Why should we bother with an alternate proof if we already know the theorem statement is true?

> *Every proof reveals much more than just the bare fact stated in the theorem, and this revelation may be more valuable than the theorem itself.*

> -L. Lovasz, Discrete Mathematics

By now, you already proved Cauchy-Schwarz by

1. Finding the closest point on a line to the origin.

2. Rewriting the *difference of the square of the norm product and the square of the dot product* as a non-negative sum (Problem Set 1).

   This time you are going to use

3. Induction

But for our induction to work, we will need to prove Cauchy-Schwarz for $n = 2$ separately.

Lemma (Cauchy-Schwarz for $n = 2$). *For real numbers x_1, x_2, y_1, y_2,*

$$|x_1 y_1 + x_2 y_2| \leq \sqrt{x_1^2 + x_2^2}\sqrt{y_1^2 + y_2^2}$$

Proof: As usual, it is easier to work with squares. Thus, we try to prove

$$(x_1 y_2 + x_2 y_2)^2 \leq (x_1^2 + x_2^2)(y_1^2 + y_2^2)$$

But this is equivalent to proving

$$0 \leq (x_1^2 + x_2^2)(y_1^2 + y_2^2) - (x_1 y_1 + x_2 y_2)^2$$

Expanding the right hand side,

$$
\begin{aligned}
(x_1^2 + x_2^2)(y_1^2 + y_2^2) - (x_1y_1 + x_2y_2)^2 &= (x_1^2y_1^2 + x_1^2y_2^2 + x_2^2y_1^2 + x_2^2y_2^2) - (x_1^2y_1^2 + 2x_1y_1x_2y_2 + x_2^2y_2^2) \\
&= \underbrace{x_1^2y_2^2}_{a^2} - \underbrace{2x_1y_1x_2y_2}_{2ab} + \underbrace{x_2^2y_1^2}_{b^2}
\end{aligned}
$$

But lo and behold, what is this? It is one of our oldest friends, the factoring:

$$(a - b)^2 = a^2 - 2ab + b^2$$

with

$$
\begin{aligned}
a &= x_1y_2 \\
b &= x_2y_1
\end{aligned}
$$

But we already know

$$(x_1y_2 - x_2y_1)^2 \geq 0$$

So Cauchy-Schwarz for the case $n = 2$ is true. $\qquad\square$

With the case $n = 2$ out of the way, let's prove the full blown Cauchy-Schwarz inequality:

Theorem (Cauchy-Schwarz Inequality). *For $x_1, x_2 \ldots x_n, y_1, y_2, \ldots y_n \in \mathbb{R}$ the following inequality always holds:*

$$|x_1y_1 + x_2y_2 \ldots + x_ny_n| \leq \sqrt{x_1^2 + x_2^2 + \ldots + x_n^2} \sqrt{y_1^2 + y_2^2 + \ldots + y_n^2}$$

Proof Summary:

- Base Case: Use property $|ab| = |a||b|$

- Inductive Step: Start with LHS and split into the sum of the n case and a_{n+1} term

- Visualize as Cauchy-Schwarz for case $n = 2$ and apply lemma.

Proof: We do induction on n to prove the property

$$P_n : |x_1y_1 + x_2y_2 \ldots + x_ny_n| \leq \sqrt{x_1^2 + x_2^2 + \ldots + x_n^2} \sqrt{y_1^2 + y_2^2 + \ldots + y_n^2}$$

is true for every positive n.

- BASE CASE, $n = 1$.

  To show

$$|x_1y_1| \leq \sqrt{x_1^2} \sqrt{y_1^2}$$

just apply basic absolute value properties:

$$|x_1 y_1| = |x_1|\,|y_1| = \sqrt{x_1^2}\sqrt{y_2^2}.$$

Thus, P_1 is true.[1]

- INDUCTIVE STEP

Assume P_n is true. We want to prove

$$P_{n+1} : |x_1 y_1 + x_2 y_2 + \ldots + x_{n+1} y_{n+1}| \le \sqrt{x_1^2 + x_2^2 + \ldots + x_{n+1}^2}\sqrt{y_1^2 + y_2^2 + \ldots + y_{n+1}^2}$$

Like with most inequalities, we start from the left and try to construct an upper bound. First, visualize the left hand side as the sum of two vectors and apply **normal**[2] triangle inequality:

$$\left| \underbrace{(x_1 y_1 + x_2 y_2 + \ldots + x_n)}_{a} + \underbrace{(x_{n+1} y_{n+1})}_{b} \right| \le \underbrace{|x_1 y_1 + x_2 y_2 + \ldots + x_n|}_{|a|} + \underbrace{|x_{n+1} y_{n+1}|}_{|b|}$$

Using the inductive hypothesis, we can bound this even further:

$$|x_1 y_1 + x_2 y_2 + \ldots + x_n| + |x_{n+1} y_{n+1}| \le \sqrt{x_1^2 + x_2^2 + \ldots x_n^2}\sqrt{y_1^2 + y_2^2 + \ldots y_n^2} + |x_{n+1} y_{n+1}|.$$

Now we do a **very nice trick**: apply the Cauchy-Schwarz inequality for the $n = 2$ case using the vectors,

$$\vec{a} = \begin{bmatrix} \sqrt{x_1^2 + x_2^2 + \ldots + x_n^2} \\ |x_{n+1}| \end{bmatrix} \qquad \vec{b} = \begin{bmatrix} \sqrt{y_1^2 + y_2^2 + \ldots + y_n^2} \\ |y_{n+1}| \end{bmatrix}$$

Cauchy-Schwarz on $\vec{a}, \vec{b}$ gives us an upper bound for the right hand side

$$\left| \underbrace{\sqrt{x_1^2 + \ldots + x_n^2}\sqrt{y_1^2 + \ldots + y_n^2} + |x_{n+1}|\,|y_{n+1}|}_{\|\vec{a}\cdot\vec{b}\|} \right| \qquad \le$$

$$\underbrace{\sqrt{\left(\sqrt{x_1^2 + \ldots + x_n^2}\right)^2 + |x_{n+1}|^2} \cdot \sqrt{\left(\sqrt{y_1^2 + \ldots + y_n^2}\right)^2 + |y_{n+1}|^2}}_{\|\vec{a}\|\,\|\vec{b}\|}$$

which, after simplifying, is

$$\sqrt{x_1^2 + x_2^2 + \ldots + x_n^2 + |x_{n+1}|^2} \cdot \sqrt{y_1^2 + y_2^2 + \ldots + y_n^2 + |y_{n+1}|^2}$$

Of course, we can drop absolute values when we square:

---

[1]Equality should be intuitive in the $n = 1$ case since any two numbers on the number line are "parallel"
[2]General triangle inequality would be circular!

$$\sqrt{x_1^2 + x_2^2 + \ldots + x_n^2 + x_{n+1}^2} \cdot \sqrt{y_1^2 + y_2^2 + \ldots + y_n^2 + y_{n+1}^2}.$$

Thus,

$$|x_1 y_1 + x_2 y_2 + \ldots + x_{n+1} y_{n+1}| \leq \sqrt{x_1^2 + x_2^2 + \ldots + x_n^2 + x_{n+1}^2} \cdot \sqrt{y_1^2 + y_2^2 + \ldots + y_n^2 + y_{n+1}^2}.$$

and the case P_{n+1} is true. This lets us conclude the Cauchy-Schwarz inequality is true. $\qquad\square$

Notice that we used induction on integers, strings, and sequences. Generally,

---

Math Mantra: Consider using induction when dealing with objects that are
constructed RECURSIVELY i.e. each successive object is expressed in terms of
the previous objects.

4.5 How Induction Should Not be Done

This is perhaps the most important memory I've collected. It is also a lie!
-Albus Dumbledore

As a teacher, I see induction incorrectly taught time and time again. In fact, I am sure your graders will complain about it after the first homework!

Here is a "proof" of Gauss' Lemma pilfered from a handout by a math teacher at a reputable local high school:

Theorem. *For any positive integer n,*

$$1 + 2 + 3 + \ldots + n = \frac{n(n+1)}{2}$$

Fake proof:

- BASE, $n = 1$

$$1 \overset{?}{=} \frac{1(1+1)}{2} = 1$$
$$1 \overset{\checkmark}{=} 1$$

- INDUCTIVE

$$1 + 2 + 3 + \ldots + n + (n+1) \overset{?}{=} \frac{(n+1)(n+2)}{2}$$

$$\frac{n(n+1)}{2} + (n+1) \overset{?}{=} \frac{(n+1)(n+2)}{2}$$

$$\frac{n^2+n}{2} + \frac{2(n+1)}{2} \overset{?}{=} \frac{(n+1)(n+2)}{2}$$

$$\frac{n^2+3n+2}{2} \overset{?}{=} \frac{(n+1)(n+2)}{2}$$

$$\frac{(n+1)(n+2)}{2} \overset{\checkmark}{=} \frac{(n+1)(n+2)}{2}$$

This mindless, mechanical process is a by-product of an education revolving around rote memorization. To us, there is some merit in the process because the kids are showing an equivalence to a tautology. But kids don't know that. Instead, they are taught:

1. Assume what you are trying to prove.

2. Manipulate to get a true statement.

This is nuttier than squirrel poo. The first step is circular: you cannot assume what you are trying to prove! Combined with the second step, it creates great evil. If this process is permitted, then you can derive anything:

$$\begin{aligned} 27232 &= 323 &\Longrightarrow \\ 0 \cdot 27232 &= 0 \cdot 323 &\Longrightarrow \\ 0 &= 0 \end{aligned}$$

By this logic, $27232 = 323$. Fail!

The better way to teach induction is to tell the student to

1. Start with some initial knowledge (the left hand side).

2. Using the given, **build towards** what you are trying to prove (the right hand side).

Starting with

$$1 + 2 + 3 + \ldots + n + (n+1) \tag{$\star$}$$

we can apply the inductive hypothesis

$$1 + 2 + 3 + \ldots + n = \frac{n(n+1)}{2}$$

to rewrite $(\star)$ as

$$\frac{n(n+1)}{2} + (n+1)$$

which is just

$$\frac{n(n+1)}{2} + \frac{2(n+1)}{2} = \frac{n^2+n+2(n+1)}{2} = \frac{(n+1)(n+2)}{2}.$$

Thus,

$$1 + 2 + 3 + \ldots + n + (n+1) = \frac{(n+1)(n+2)}{2}$$

Building *from* what we know *to* what we want to conclude is a natural process. The natural process of **REASONING**. Anyways, that's my rant on high school induction. Thanks for listening.

4.6 Under-determined Systems Lemma

With induction at our disposal, we can prove the Under-determined Systems Lemma. But what is this lemma?

In Algebra II, you learned that a system of equations can be visualized as an intersection of lines. In the case of

$$\begin{aligned} 2x &+ 2y &= 0 \\ 3x &+ 4y &= 0 \end{aligned}$$

there must only be **one** solution since these lines are non-parallel. Also note that this one solution must be $(0,0)$ since this is a **homogeneous system**.

However, suppose we have only one equation:

$$2x + 2y = 0.$$

Then there exists a solution that is not $(0,0)$. In fact, we have infinitely many solutions, since any point on the line

$$y = -x$$

satisfies the equation.

The Under-determined Systems Lemma generalizes this idea. Namely, in a homogeneous system with more equations than unknowns, you will always have a non-trivial solution. This is very intuitive: just think of equations as *constraints* and unknowns as *freedoms*. You are guaranteed a solution when you have more freedom than constraint.

Theorem (Under-determined Systems Lemma). *Any homogeneous system of m equations with more than m unknowns has a non-trivial solution.*

Proof: We proceed by induction on the number of equations, k, to prove the property

> P_k : *Any homogeneous system with k equations and more than k unknowns has at least one non-trivial solution.*

holds for every positive integer k.

- BASE CASE, $k = 1$.

Consider a system of 1 equation with n unknowns, where $n > k$:

$$a_{11}x_1 + a_{12}x_2 + \ldots + a_{1n}x_n = 0$$

We have two possibilities:

$$a_{11} = 0 \quad \text{or} \quad a_{11} \neq 0$$

- If $a_{11} = 0$, we can form a non-trivial solution by simply setting the first unknown to 1 and the rest to 0:

$$
\begin{aligned}
x_1 &= 1 \\
x_2 &= 0 \\
&\vdots \\
x_n &= 0
\end{aligned}
$$

- If $a_{11} \neq 0$, then set all unknowns except the first to 1:

$$
\begin{aligned}
x_2 &= 1 \\
x_3 &= 1 \\
&\vdots \\
x_n &= 1
\end{aligned}
$$

Now solve for x_1:

$$
\begin{aligned}
a_{11}x_1 + a_{12}(1) + a_{13}(1) + \ldots + a_{1n}(1) &= 0 & \Longrightarrow \\
a_{11}x_1 &= -(a_{12} + a_{13} + \ldots + a_{1n}) & \Longrightarrow \\
x_1 &= \frac{-(a_{12} + a_{13} + \ldots + a_{1n})}{a_{11}}
\end{aligned}
$$

This means

$$
\begin{aligned}
x_1 &= \frac{-(a_{12} + a_{13}\ldots + a_{1n})}{a_{11}} \\
x_2 &= 1 \\
x_3 &= 1 \\
&\vdots \\
x_n &= 1
\end{aligned}
$$

is a non-trivial solution.

In either case, we have a non-trivial solution. Thus, the base case, P_1, is true.

- INDUCTIVE STEP

Assume P_m is true. We want to show P_{m+1} is true, i.e.

P_{m+1} : *Any homogeneous system with $m + 1$ equations and more than $m + 1$ unknowns has at least one non-trivial solution.*

Consider an arbitrary homogeneous system with $m + 1$ equations and n unknowns, where $n > m + 1$:

$$
\begin{aligned}
a_{11}x_1 + a_{12}x_2 + a_{13}x_3 + \ldots + a_{1n}x_n &= 0 \\
a_{21}x_1 + a_{22}x_2 + a_{23}x_3 + \ldots + a_{2n}x_n &= 0 \\
\vdots \qquad\quad \vdots \qquad\quad \vdots \qquad\quad \vdots \qquad\quad \vdots \\
a_{(m+1)1}x_1 + a_{(m+1)2}x_2 + a_{(m+1)3}x_3 + \ldots + a_{(m+1)n}x_n &= 0
\end{aligned}
$$

The first step of Gaussian Elimination *preserves* the solutions (we will prove this later)! Thus, we can perform this first step to transform the system into one of two possible cases:

- CASE 1

The system reduces to

$$
\begin{aligned}
0x_1 + a'_{12}x_2 + a'_{13}x_3 + \ldots + a'_{1n}x_n &= 0 \\
0x_1 + a'_{22}x_2 + a'_{23}x_3 + \ldots + a'_{2n}x_n &= 0 \\
\vdots \qquad\quad \vdots \qquad\quad \vdots \qquad\quad \vdots \qquad\quad \vdots \\
0x_1 + a'_{(m+1)2}x_2 + a'_{(m+1)3}x_3 + \ldots + a'_{(m+1)n}x_n &= 0
\end{aligned}
$$

This means that the value of x_1 **does not matter**. Setting

$$
\begin{aligned}
x_1 &= 1 \\
x_2 &= 0 \\
&\vdots \\
x_n &= 0
\end{aligned}
$$

yields a non-trivial solution.

- CASE 2

The system reduces to

$$
\begin{aligned}
1x_1 + a'_{12}x_2 + a'_{13}x_3 + \ldots + a'_{1n}x_n &= 0 \\
0x_1 + a'_{22}x_2 + a'_{23}x_3 + \ldots + a'_{2n}x_n &= 0 \\
\vdots \qquad\quad \vdots \qquad\quad \vdots \qquad\quad \vdots \qquad\quad \vdots \\
0x_1 + a'_{(m+1)2}x_2 + a'_{(m+1)3}x_3 + \ldots + a'_{(m+1)n}x_n &= 0
\end{aligned}
$$

But notice that, if we ignore the first equation, we have a system of m equations with $n - 1$ unknowns:

$$
\begin{aligned}
x_1 + a'_{12}x_2 + a'_{13}x_3 + \ldots + a'_{1n}x_n &= 0 \\
a'_{22}x_2 + a'_{23}x_3 + \ldots + a'_{2n}x_n &= 0 \\
\vdots \qquad\quad \vdots \qquad\quad \vdots \qquad\quad \vdots \\
a'_{(m+1)2}x_2 + a'_{(m+1)3}x_3 + \ldots + a'_{(m+1)n}x_n &= 0
\end{aligned}
$$

By applying the inductive hypothesis on this subsystem, we get a non-trivial solution:

$$
\begin{aligned}
x_2 &= s_2 \\
x_3 &= s_3 \\
&\vdots \\
x_n &= s_n
\end{aligned}
$$

Now plug this solution into the first equation

$$1x_1 + a_{12}'s_2 + a_{13}'s_3 + \ldots + a_{1n}'s_n = 0$$

and solve for x_1:

$$x_1 = -(a_{12}'s_2 + a_{13}'s_3 + \ldots + a_{1n}'s_n)$$

Thus,

$$
\begin{aligned}
x_1 &= -(a_{12}'s_2 + a_{13}'s_3 + \ldots + a_{1n}'s_n) \\
x_2 &= s_2 \\
&\vdots \\
x_n &= s_n
\end{aligned}
$$

is a non-trivial solution of the larger system. $\square$

By the way, you should ask yourself,

Why should we care about the Under-determined Systems Lemma?

One reason is that this lemma is used to prove the Linear Dependence Lemma. And without the Linear Dependence Lemma, a lot of our upcoming subspace theory (like basis and dimension), wouldn't even make sense!

4.7   Linear Dependence Lemma

Consider the question: can you choose 6 numbers from the set

$$\{1, 2, 3, 4, 5, 6, 7, 8, 9, 10\}$$

such that none are even? The answer is obviously **no** since the set only contains 5 odd numbers.

The Linear Dependence Lemma follows the same pigeon-hole philosophy: you cannot choose $k + 1$ vectors from the set

$$\text{span}\{\vec{v_1}, \vec{v_2}, \ldots, \vec{v_k}\}$$

such that these vectors are linearly independent.

At first sight, the proof of the Linear Dependence Lemma looks difficult. It's not. The only reason it seems difficult is that it requires a healthy dose of *book-keeping*.

> ```
> Math Mantra:  Don't be deceived by cumbersome notation.  Instead, look for the
>                              BIG PICTURE!
> ```

Here is the big picture:

To prove linear dependence, you want to find a non-trivial solution (i.e. not $x_1 = \ldots = x_{k+1} = 0$) to some equation of the form

$$x_1 \vec{w}_1 + x_2 \vec{w}_2 + \ldots + x_k \vec{w}_k + x_{k+1} \vec{w}_{k+1} = \vec{0}.$$

By substituting the definition of $\vec{w}_i$, we can rewrite this equation as

$$c_1 \vec{v}_1 + c_2 \vec{v}_2 + \ldots + c_k \vec{v}_k = \vec{0}$$

where each coefficient c_i is a *function* of the unknowns $x_1, x_2, \ldots, x_{k+1}$. To find some solution to this equation, it suffices to find $x_1, x_2, \ldots, x_{k+1}$ that makes each coefficient zero for then we have:

$$0\vec{v}_1 + 0\vec{v}_2 + \ldots + 0\vec{v}_k = \vec{0}.$$

Therefore, we will solve the system

$$
\begin{aligned}
c_1 &= 0 \\
c_2 &= 0 \\
&\vdots \\
c_k &= 0
\end{aligned}
$$

When we write out the coefficients $c_1, c_2, \ldots, c_k$ this is a homogeneous system of equations with more unknowns than equations. Just apply Under-determined Systems Lemma and we're done!

Theorem (Linear Dependence Lemma). *Let $\vec{v}_1, \vec{v}_2, \ldots, \vec{v}_k \in \mathbb{R}^n$. Any set of $k+1$ vectors taken from*

$$\text{span}\{\vec{v}_1, \vec{v}_2, \ldots, \vec{v}_k\}$$

is linearly dependent.

Proof Summary:

- We want to find a non-trivial solution to

$$x_1 \vec{w}_1 + x_2 \vec{w}_2 + \ldots + x_{k+1} \vec{w}_{k+1} = \vec{0}.$$

- Rewrite each $\vec{w}_i$ as a linear combination of $\vec{v}_1, \vec{v}_2, \ldots, \vec{v}_k$.

- Rewrite the new equation as a sum of scaled $\vec{v}_i$.

- Set each coefficient of $\vec{v}_i$ to 0.

- This is a homogeneous system with k equations, $k+1$ unknowns: apply Under-determined Systems Lemma.

Proof: Let $\vec{w}_1, \vec{w}_2, \ldots, \vec{w}_{k+1}$ be arbitrary vectors such that

$$\vec{w}_1, \vec{w}_2, \ldots, \vec{w}_{k+1} \in \operatorname{span}\{\vec{v}_1, \vec{v}_2, \ldots, \vec{v}_k\}$$

Since we want to prove these vectors are linearly dependent, by definition, we need a non-trivial solution to

$$x_1\vec{w}_1 + x_2\vec{w}_2 + \ldots + x_{k+1}\vec{w}_{k+1} = \vec{0}. \qquad (\star)$$

First, expand the definition of the w_i i.e, write each of them as a member of the span. We *could* try

$$
\begin{aligned}
\vec{w}_1 &= \alpha_1\vec{v}_1 &+ \alpha_2\vec{v}_2 &+ \ldots &+ \alpha_k\vec{v}_k \\
\vec{w}_2 &= \alpha_1\vec{v}_1 &+ \alpha_2\vec{v}_2 &+ \ldots &+ \alpha_k\vec{v}_k \\
&\vdots \\
\vec{w}_{k+1} &= \alpha_1\vec{v}_1 &+ \alpha_2\vec{v}_2 &+ \ldots &+ \alpha_k\vec{v}_k
\end{aligned}
$$

But this is a **bone-headed** labelling. Single variable notation's not going to cut it!

Why? We are going to have overlapping variables. Namely the above says that

$$\vec{w}_1 = \vec{w}_2 = \ldots = \vec{w}_{k+1}$$

This is not our intent! A smarter idea is to order our variables using double subscripts:

$$
\begin{aligned}
\vec{w}_1 &= \alpha_{11}\vec{v}_1 &+ \alpha_{12}\vec{v}_2 &+ \ldots &+ \alpha_{1k}\vec{v}_k \\
\vec{w}_2 &= \alpha_{21}\vec{v}_1 &+ \alpha_{22}\vec{v}_2 &+ \ldots &+ \alpha_{2k}\vec{v}_k \\
&\vdots \\
\vec{w}_{k+1} &= \alpha_{(k+1)1}\vec{v}_1 &+ \alpha_{(k+1)2}\vec{v}_2 &+ \ldots &+ \alpha_{(k+1)k}\vec{v}_k
\end{aligned}
$$

Much better! Now, it resembles the coefficients of a system of equations (the wheel starts to turn)!

Substituting $\vec{w}_i$ into $(\star)$ yields

$$x_1\underbrace{\left(\alpha_{11}\vec{v}_1 + \ldots + \alpha_{1k}\vec{v}_k\right)}_{\vec{w}_1} + x_2\underbrace{\left(\alpha_{21}\vec{v}_1 + \ldots + \alpha_{2k}\vec{v}_k\right)}_{\vec{w}_2} + \ldots + x_{k+1}\underbrace{\left(\alpha_{(k+1)1}\vec{v}_1 + \ldots + \alpha_{(k+1)k}\vec{v}_k\right)}_{\vec{w}_{k+1}} = 0$$

Distributing, we have

$$\left(x_1\alpha_{11}\vec{v}_1 + \ldots + x_1\alpha_{1k}\vec{v}_k\right) + \left(x_2\alpha_{21}\vec{v}_1 + \ldots + x_2\alpha_{2k}\vec{v}_k\right) + \ldots + \left(x_{k+1}\alpha_{(k+1)1}\vec{v}_1 + \ldots + x_{k+1}\alpha_{(k+1)k}\vec{v}_k\right) = \vec{0}$$

Now, group the $\vec{v}_i$ terms:

$$\left(x_1\alpha_{11} + x_2\alpha_{21} + \ldots + x_{k+1}\alpha_{(k+1)1}\right)\vec{v}_1 + \ldots + \left(x_1\alpha_{1k} + x_2\alpha_{2k} + \ldots + x_{k+1}\alpha_{(k+1)k}\right)\vec{v}_k = \vec{0}$$

To solve this equation, it suffices to find $x_1, x_2, \ldots, x_n$ such that the *coefficient* of each of the $\vec{v}_i$ are zero. This gives us the system

$$
\begin{aligned}
x_1\alpha_{11} &+ x_2\alpha_{21} &+ \ldots &+ x_{k+1}\alpha_{(k+1)1} &= 0 \\
x_1\alpha_{12} &+ x_2\alpha_{22} &+ \ldots &+ x_{k+1}\alpha_{(k+1)2} &= 0 \\
\vdots &\quad &\vdots \quad &\vdots &\quad \vdots \\
x_1\alpha_{1k} &+ x_2\alpha_{2k} &+ \ldots &+ x_{k+1}\alpha_{(k+1)k} &= 0
\end{aligned}
$$

But this system has k equations with $k+1$ unknowns! So by the Under-determined Systems Lemma, we have a non-trivial solution. Awesome. $\square$

Lecture 5

Keeping it Real

One must be able to say at all times: instead of points, straight lines, and planes-tables, chairs, and beer mugs.

-David Hilbert

Goals: After introducing a series of axioms, we will use the field and ordering properties to rigorously derive real number theorems. In particular, we state the Completeness Axiom. This *fundamental* axiom will be used multiple times throughout the course. We also discuss how to prove *existence* and *uniqueness*.

5.1 Thinking Axiomatically

There is nothing more important to a mathematician than a proof. But,

What is a proof?

Without going too much into philosophy, we can say that a proof is a *truth-preserving operation*. It takes some collection of true statements and deduces a new true statement.

Naturally, we can try to trace all true statements back to some *initial set* of truths which we call *axioms*.

The process of taking some theory and formally identifying the axioms at its foundation is known as *axiomization*. Particularly, we would like to axiomatize the **theory of real numbers**. This axiomization is going to be in three parts:

- **The Field Axioms**

- **The Ordering Axioms**

- **The Completeness Axiom**

For the first item, we are going to introduce the *abstract notions* of a *group* and a *field*. Then, to check that our *axioms make sense*, we are going to apply *only* the *algebraic properties* of a field to derive real number properties. This means *forgetting* all the meaning behind the symbols

$$+, -, \times, \div$$

89

and *forgetting* all the real numbers

$$2, \pi, 7, \ldots$$

For all we care, we could call these numbers tables, chairs, and beer mugs. We will treat numbers as purely *algebraic objects* to which we can apply formal *algebraic rules*. For example, consider the object

$$-1 \cdot (2 + 3).$$

We **do not know** that this is -5. However, we can formally use the rule

$$-1 \cdot (x + y) = -1 \cdot x + -1 \cdot y$$

to derive

$$-1 \cdot (\underbrace{2}_{x} + \underbrace{-3}_{y}) = -1 \cdot \underbrace{2}_{x} + -1 \cdot \underbrace{3}_{y}.$$

But before I cast *Obliviate*, we start off with a discussion of one of the most fundamental concepts in all mathematics: *uniqueness*.

5.2   Proof Technique: Uniqueness

> *In the end, there can be only one!*
>
> -Duncan McCloud

When you walk into a bar, you may see a strange yellow bottle:

This bottle holds Galliano, a sugary vanilla-flavored liqueur. If you ever ask a bartender for a drink made with Galliano he will *always* make a Harvey Wallbanger. This is because

A Harvey Wallbanger is the only[1] drink in the entire universe that uses Galliano.

We have *uniquely identified*

Harvey Wallbanger

with the property

---

[1] As far as I know, but we will assume this for mathematical purposes.

made with Galliano.

Generally, we say that

Object x uniquely satisfies a property P

if x is the *only* object that satisfies property P. Technically, this is expressed as: x satisfies P and for any object y that satisfies P, it *must* be the case $y = x$.

As a more mathematical example, consider the roots of

$$\sin(x).$$

There are infinitely many roots, namely

$$x = n\pi$$

for any integer n. But there is only one root in the interval $\left[-\frac{\pi}{2}, \frac{\pi}{2}\right]$, namely, $x = 0$. Therefore, we have *uniquely identified*

$$\boldsymbol{x = 0}$$

as

The root of $\sin(x)$ **in interval** $\left[-\frac{\pi}{2}, \frac{\pi}{2}\right]$.

Two reasons why we study uniqueness is that

- **Uniqueness is important in the study of functions.**
  By *definition*, a function has a *unique* output for every input. Moreover, for a function to have an *inverse*, every point in the image must be mapped from a *unique* point in the domain.

- **Uniqueness can be used in a proof by contradiction.**
  If an object has a unique representation of a certain form and we have two different ways to express it in that form, then we have a contradiction. Here's a fun example:

Example. $\sqrt[5]{3}$ *is irrational.*

Proof Summary:

- Suppose rational.

- Rewrite as $3q^5 = p^5$.

- The power of 3 in the prime factorization of p^5 is a multiple of 5.

- The power of 3 in the prime factorization of $3q^5$ is one more than a multiple of 5.

- Contradicts uniqueness of prime factorization.

Proof: Suppose $\sqrt[5]{3}$ is rational. Then

$$\sqrt[5]{3} = \frac{p}{q}$$

for integers p, q. Raising both sides to the fifth power gives

$$3 = \frac{p^5}{q^5}$$

so

$$3q^5 = p^5.$$

In the prime factorization of

$$p^5,$$

all the prime powers must be multiples of 5. This is because we can write p in terms of its prime factorization,

$$p = (p_1^{\alpha_1} p_2^{\alpha_2} \ldots p_r^{\alpha_r})^5$$

and expand

$$p^5 = \underbrace{(p_1^{\alpha_1} p_2^{\alpha_2} \ldots p_r^{\alpha_r})}_{p}{}^5 = p_1^{5\alpha_1} p_2^{5\alpha_2} \ldots p_r^{5\alpha_r}$$

Since 3 divides p^5 (from above),

> *The **power of** 3 in the prime factorization of p^5 is a multiple of 5.*

Likewise, we can show the prime factorization of q^5 only includes multiples of 5:

$$q^5 = \underbrace{(q_1^{\beta_1} q_2^{\beta_2} \ldots q_s^{\beta_s})}_{q}{}^5 = q_1^{5\beta_1} q_2^{5\beta_2} \ldots q_s^{5\beta_s}.$$

Therefore,

> *The **power of** 3 in the prime factorization of $3q^5$ is **one more** than a multiple of 5.*

But

$$3q^5 = p^5$$

so by **uniqueness of prime factorization** the power of 3 in the unique prime factorization of $3q^5$ must be both

- A multiple of 5.

- One more than a multiple of 5.

which is absurd. Thus, $\sqrt[5]{3}$ is irrational.   □

Uniqueness is important. But how do we prove it?

Simple:

> **To prove uniqueness, consider two objects that satisfy the same property. Using only the fact that they satisfy this property, prove that the two objects *must* be equal.**

As a first example, recall the Algebra II topic of inverse functions. To construct an inverse, you

1. "Switched" x and y.

2. Solved for y in terms of x

However, mindlessly applying this procedure doesn't *guarantee*[1] *an inverse exists*. Namely, we need to first prove the function is *injective*. This means that every output is mapped from a *unique* input.[2]

Example. *Consider the function*
$$f(x) = 3x + 5$$
Then every point in its image is mapped from a unique point in the domain.

Proof: Suppose we have two points a, b in the domain that are mapped to the same image point:
$$f(a) \ = \ f(b)$$
By definition,
$$3a + 5 = 3b + 5.$$
which implies
$$a = b.$$
Thus, if we have two points that are mapped to the same image point, then those two points must be equal. $\qquad\square$

The prototypical example is proving the uniqueness of the *Division Algorithm*:

In elementary school, you performed long division using a *remainder term*. For example, when you divided 22 by 4, the quotient was 5 with remainder 2:

$$\begin{array}{r} 5R2 \\ 4\overline{\smash{\big)}\ 22} \\ \underline{20} \\ 2 \end{array}$$

Now that we are adults, the proper way to express this division is
$$22 = 4q + r$$
where
$$\begin{aligned} q &= 5 \\ r &= 2 \end{aligned}$$

---

[1]We delay the proof that injectivity implies that the inverse function exists: see **Lecture 34**.

[2]As an analogy, think of Cryptography. Suppose two messages A and B were encoded as the same coded message C. When we receive C and try to decode (invert) it, we have no idea whether the original message was A or B!

I claim that when you divide an integer by any positive integer, the quotient and remainder pair[1] (q, r) is always unique.

Example (Division Algorithm). *Given a positive integer b, every integer n can be written uniquely in the form*

$$n = bq + r$$

where q and r are integers, and r is either $0, 1, 2 \dots, b - 1$. Formally, if we have both

$$n = bq + r$$
$$n = bq' + r'$$

satisfying the conditions above, then

$$(q, r) = (q', r')$$

Proof Summary:

- Consider two different ways to write $n = bq + r$.

- Isolate q's on one side and r's on other.

- Absolute value both sides.

- Suppose q's are different. We have a number both greater than or equal to b and less than b, contradiction. Conclude $q' = q$.

- Substitute $q' = q$ to immediately get $r' = r$.

- Conclude $(q, r) = (q', r')$.

Proof: Suppose

$$n = bq + r$$
$$n = bq' + r'$$

for integers q, q', r, r' with

$$0 \leq r, r' \leq b - 1.$$

Equating the two right hand expressions and isolating the $r's$ on one side

$$bq + r = bq' + r'.$$

In other words,

$$b(q - q') = r - r'.$$

Suppose $q \neq q'$. Because we don't want to be bothered with negatives, take absolute values of both sides.

$$|b(q - q')| = |r - r'|$$

---

[1]Two pairs are different if they differ in *at least one component*. For example, (2,3) is different from the pairs (3,5), (2,4), and (4,3).

and use the absolute value property $|xy| = |x||y|$ to split the (LHS)

$$|b|\,|q - q'| = |r - r'|$$

Because b is positive, we drop the absolute value:

$$b\,|q - q'| = |r - r'| \qquad (\star)$$

Since $q - q' \neq 0$, we know $|q - q'|$ is some **positive integer** and therefore

$$b|q - q'| \geq b \cdot \underbrace{1}_{} = b$$

However,

$$|r - r'| \leq b - 1.$$

This is because $|r - r'|$ is biggest if one of r, r' is 0 and the other is $b - 1$. Applying these bounds to $(\star)$,

$$\underbrace{b\,|q - q'|}_{\geq b} = \underbrace{|r - r'|}_{<b},$$

which is impossible! Therefore

$$q = q'$$

Substituting

$$bq + r = b\underbrace{q}_{q'} + r',$$

we solve for

$$r = r'.$$

In conclusion, if

$$\begin{aligned} n &= bq + r \\ n &= bq' + r' \end{aligned}$$

for integers q, q', r, r' with

$$0 \leq r, r' \leq b - 1,$$

it must be the case that

$$(q, r) = (q', r'). \qquad \square$$

For the next example of uniqueness, consider the question

Is there any function other than e^x, that goes through $(0, 1)$ and is its own derivative?

The answer is **no.** To prove this, we use one of my favorite tricks in mathematics:

```
Math Mantra:   If you want to prove that a differentiable function g is uniquely
                                                                  1
            some (non-zero) function f, try differentiating g(x) · ―――― .
                                                                 f(x)
```

In **Lecture 3**, we proved that

f′ is the zero function if and only if f is a constant.

Therefore,

$$\left(g(x) \cdot \frac{1}{f(x)}\right)' = 0$$

implies

$$g(x) \cdot \frac{1}{f(x)} = k$$

for some constant k. Hence,

$$g(x) = kf(x).$$

To solve for k, we can plug in any number we want into x e.g. 0:

$$k = \frac{g(0)}{f(0)}.$$

If $k = 1$, then

$$g(x) = f(x)$$

as needed.

Example. *The **unique function** that is its own derivative and has value 1 at $x = 0$ is*

$$f(x) = e^x.$$

Proof Summary:

- Consider u that satisfies the same property.

- Differentiate $u(x)e^{-x}$

- Derivative is 0, so $u(x)e^{-x}$ is constant.

- Plug in 0 to solve for that constant. Conclude $u(x) = e^x$.

Proof: Suppose there is some other function u that has the property

$$\begin{aligned} u'(x) &= u(x) \\ u(0) &= 1 \end{aligned}$$

Consider

$$u(x) \cdot \frac{1}{e^x}$$

which is equivalently

$$u(x)e^{-x}.$$

Differentiate according to the product rule:

$$(u(x)e^{-x})' = u'(x)e^{-x} - u(x)e^{-x}.$$

But $u'(x)$ is **its own derivative**, so this expression simplifies to

$$(u(x)e^{-x})' = u'(x)e^{-x} - \underbrace{u'(x)}_{u(x)}e^{-x} = 0$$

Therefore,

$$u(x)e^{-x} = k$$

for some constant k. Since we know $u(0) = 1$, plug in 0 to solve for k:

$$u(0)e^0 = 1.$$

Therefore $k = 1$ and

$$u(x)e^{-x} = 1$$

Multiplying by e^{-x}, we get

$$u(x) = e^x$$

and conclude that e^x is the unique function that is its own derivative and value 1 at $x = 0$. $\qquad\square$

5.3 Abelian Groups

> *What's purple and commutes?*
> *An Abelian grape.*

To understand the field axioms, first we need to understand *abelian groups*.

Consider some set S and some function $\star$ that takes any two inputs

$$x, y \in S$$

and returns an output

$$\star(x, y) \in S$$

Alternatively, we write the output as

$$x \star y.$$

Then, we call $\star$ a **binary function on** S. In general,

Definition. *For set S and binary function $\star : S \times S \to S$, we call the pair $(S, \star)$ an* **abelian group** *if the following properties hold:*

- **Existence of an identity element**
  There is some element in the set such that when you apply the $\star$ operation, you get the other input: there exists $e \in S$ such that
  $$e \star x = x \star e = x$$
  for all $x \in S$.

- **Existence of inverses**
  For any input x, you can find some other input such that applying the $\star$ operation outputs the identity element: for all $x \in S$, there exists[1] $x^\dagger \in S$ such that
  $$x \star x^\dagger = x^\dagger \star x = e.$$

- **Commutativity**
  The order in which the $\star$ operation is applied does not matter: for all $x, y \in S$,
  $$x \star y = y \star x.$$

- **Associativity**
  Applying $\star$ to x and $y \star z$ is the same as applying $\star$ to $x \star y$ and z: for all $x, y, z \in S$,
  $$x \star (y \star z) = (x \star y) \star z.$$

Here are a few fundamental properties you should know about abelian groups:

Theorem. *The identity element e is unique.*

Proof: Suppose there are two identity elements e_1 and e_2. By definition of e_1 being an identity, for any element $x \in S$,
$$e_1 \star x = x.$$

Taking $x = e_2$ gives
$$e_1 \star \underbrace{e_2}_{x} = \underbrace{e_2}_{x}. \tag{1}$$

Likewise, we know e_2 is an identity element, so
$$x \star e_2 = x$$

---

[1] Notice that element $x^\dagger$ depends on the choice of x. In other words, the inverse is a *function* of x.

for any $x \in S$. In particular,

$$\underbrace{e_1}_{x} \star e_2 = \underbrace{e_1}_{x} \qquad (2)$$

Combining (1) and (2),

$$e_1 = e_1 \star e_2 = e_2. \qquad \square$$

Theorem. *For each $x \in S$, the inverse of x is unique.*

Proof: Suppose x has two inverses, say i_1 and i_2. Then we know

$$
\begin{aligned}
i_1 &= i_1 \star e & \text{(Identity)} \\
&= i_1 \star (x \star i_2) & (i_2 \text{ is an inverse}) \\
&= (i_1 \star x) \star i_2 & \text{(Associativity)} \\
&= e \star i_2 & (i_1 \text{ is an inverse}) \\
&= i_2 & \text{(Identity)}
\end{aligned}
$$

$\square$

Theorem. *The inverse of the inverse of an element is the original element:*

$$x = \left(x^\dagger\right)^\dagger.$$

Proof: Since inverses are unique by the last theorem, proving

$$x = \left(x^\dagger\right)^\dagger$$

means we have to show x is the inverse of $x^\dagger$ i.e.

$$x \star x^\dagger = x^\dagger \star x = e.$$

But this is immediately true by the definition of $x^\dagger$. $\square$

5.4 Fields

To form a *field*, we start with an abelian group

$$(F, \bullet)$$

with identity element e_1. Then we form another abelian group on F except with the identity e_1 *removed* (you'll see why):

$$(F - \{e_1\}, \star)$$

Finally, we relate the two operations via a *distributive law*. Formally,

Definition. *For a set F and binary functions $\bullet, \star$, we say that $(F, \bullet, \star)$ is a **field** if*

- *$(F, \bullet)$ is an abelian group with identity e_1.*

- *$(F - \{e_1\}, \star)$ is an abelian group with identity e_2.*

- *The identity elements of the two groups are different: $e_1 \neq e_2$.*

- *$\star, \bullet$ satisfy the distributive law: for any $a, b, c \in F$*

$$a \star (b \bullet c) = (a \star b) \bullet (a \star c)$$

Note that we take e_1 away from F to get an abelian group under $\star$ and require

$$e_1 \neq e_2.$$

This is because we **do not** want $\bullet, \star$ to be *the same* abelian group operation. Moreover, this rule *forbids* the existence of an inverse of e_1 under $\star$. To see why, first notice

Theorem. *For any $x \in F$,*

$$e_1 \star x = e_1.$$

Proof: For any $x \in F$,

$$
\begin{aligned}
e_1 \star x &= (e_1 \bullet e_1) \star x && \text{(Identity)} \\
&= x \star (e_1 \bullet e_1) && \text{(Commutativity)} \\
&= (x \star e_1) \bullet (x \star e_1) && \text{(Distributive Law)} \\
&= (e_1 \star x) \bullet (e_1 \star x) && \text{(Commutativity)}
\end{aligned}
$$

Applying the inverse of $e_1 \star x$ (under $\bullet$) to both sides of

$$e_1 \star x = (e_1 \star x) \bullet (e_1 \star x)$$

yields

$$e_1 = e_1 \star x \qquad\qquad \square$$

Using this theorem, we can show

Theorem. *The inverse of e_1 under $\star$ **does not exist**.*

Proof: Suppose it does exist. Let $e_1^{\dagger}$ denote the inverse of e_1 under $\star$.

By definition of inverse (under $\star$),

$$e_1 \star e_1^\dagger = e_2.$$

Yet the previous theorem gives us

$$e_1 \star e_1^\dagger = e_1.$$

Therefore,

$$e_1 = e_2$$

which we already forbid! $\qquad\square$

5.5 Field Axiom

The first axiom[1] that *we accept* about the reals is

<u>FIELD AXIOM</u>

$\mathbb{R}$ *is a field.*

In particular,

- $(\mathbb{R}, +)$ is an abelian group with identity 0. The inverse of x under $+$ is denoted by $-x$.

- $(\mathbb{R} - \{0\}, \times)$ is an abelian group with identity 1. The inverse of x under $\times$ is denoted by x^{-1}.

Because we have already proven results about general groups and fields, these theorems apply *in particular* to $\mathbb{R}$. Generally,

---

Math Mantra: If a theorem about an object is proven from certain properties of that object, then the theorem also holds true for any OTHER object that satisfies the same properties.

Therefore,

GENERAL GROUP THEOREM	$(\mathbb{R}, +)$	$(\mathbb{R} - \{0\}, \times)$
The identity is unique	0 is unique	1 is unique
The inverse is unique	$-x$ is unique	x^{-1} is unique
The inverse of the inverse is the original element	$-(-x) = x$	$\left(x^{-1}\right)^{-1} = x$

Moreover, the field theorems give us

$$0 \times x = 0 \ \textit{for every } x \in \mathbb{R}$$

and

---

[1]In most texts, this is introduced as a *set* of axioms.

$$0^{-1} \text{ does not exist.}$$

In fact, by **only applying the properties of a field**, we can derive several fundamental facts about the reals.

There are a few reasons why we focus on using *only* the axioms:

- **To convince ourselves that our axioms aren't bone-headed.**
  If we can use our axioms to prove a property that does not actually hold for the reals, then we have made a *lousy* choice of axioms.

- **To generalize.**
  Properties derived from the field axioms hold for *any* field. In particular, it applies[1] to $\mathbb{Q}$ and $\mathbb{C}$.

- **To design Logic Systems.**
  We can model the real numbers in a *Logic System*.[2] In such a system, we treat numbers as purely *syntax* and apply formal *syntax manipulation rules*. Moreover, we can use such a system to program a *proof-solver*.

Using only the field axioms (and their derived properties), we can prove the following fundamental results for $\mathbb{R}$. Note that these results rely on the *distributive law*.[3]

Theorem. *For any $a, b \in \mathbb{R}$ with $b \neq 0$,*

$$-(a \times b^{-1}) = (-a) \times b^{-1}$$

Proof: By definition of "$-$", we want to show $(-a) \times b^{-1}$ is the additive inverse of $a \times b^{-1}$:

$$(-a) \times b^{-1} + a \times b^{-1} = 0$$

But we can directly check that

$$
\begin{aligned}
(-a) \times b^{-1} + a \times b^{-1} &= b^{-1} \times (-a) + b^{-1} \times a & \text{(Commutativity)} \\
&= b^{-1} \times (-a + a) & \text{(Distributive Law)} \\
&= b^{-1} \times 0 & \text{(Additive Inverse)} \\
&= 0 \times b^{-1} & \text{(Commutativity)} \\
&= 0 & (0 \times a = 0)
\end{aligned}
$$

Thus, $-a \times b^{-1}$ is the additive inverse of $a \times b^{-1}$. $\square$

---

[1] My favorite field is $\mathbb{F}_{scho}$.

[2] To learn more, I highly recommend taking **Phil 151**: Professor Sommer is an *excellent* lecturer.

[3] Typically, if you are going to prove a result involving *both* operations, it is going to involve the distributive law. This is because it's the only field property *relating* both operations.

Theorem. *For any $a \in \mathbb{R}$*

$$-1 \times a = -a.$$

Proof: The above statement really means that we want to show $-1 \times a$ is the additive inverse of a. Precisely,

$$(-1 \times a) + a = 0$$

But we can see

$$
\begin{aligned}
(-1 \times a) + a &= -1 \times a + 1 \times a && \text{(Multiplicative Identity)} \\
&= a \times -1 + a \times 1 && \text{(Commutativity)} \\
&= a \times (-1 + 1) && \text{(Distributive Law)} \\
&= (-1 + 1) \times a && \text{(Commutativity)} \\
&= 0 \times a && \text{(Additive Inverse)} \\
&= 0 && (0 \times a = 0)
\end{aligned}
$$

Thus, we can conclude

$$-1 \times a = -a. \qquad \square$$

Theorem. *In $\mathbb{R}$,*

$$-1 \times -1 = 1$$

Proof: Plugging $a = -1$ into the preceding theorem yields

$$-1 \times -1 = -(-1)$$

But we know the inverse of an inverse is the original element, so

$$-1 \times -1 = 1. \qquad \square$$

Theorem. *If $a \times b = 0$ then either[1] $a = 0$ or $b = 0$.*

Proof: Let $a \times b = 0$ and suppose it is not the case that either $a = 0$ or $b = 0$. Then $a \neq 0$ and $b \neq 0$. In particular, a has a multiplicative inverse a^{-1}, so

$$
\begin{aligned}
a \times b &= 0 && \implies \\
a^{-1} \times (a \times b) &= a^{-1} \times 0 && \implies \\
(a^{-1} \times a) \times b &= a^{-1} \times 0 && \implies \\
b &= 0
\end{aligned}
$$

a contradiction since $b \neq 0$. Thus, it must be the case that either $a = 0$ or $b = 0$. $\qquad \square$

---

[1]This does not preclude $a = 0$ **and** $b = 0$.

5.6   Shorthand Notations

You may be wondering whether we need to define axioms for *subtraction* and *division*. There's no need: we've already done the work when we defined inverses. Subtraction and division are really just shorthand notations for "combine with the inverse." Specifically, we define

$$a - b$$

to be shorthand for

$$a + (-b)$$

and

$$\frac{a}{b}$$

to be shorthand for

$$a \times b^{-1}$$

Here is a list of shorthands involving $+, \times$:

Shorthand	Definition
$a - b$	$a + (-b)$
$\dfrac{a}{b}$	$a \times b^{-1}$
ab	$a \times b$

and here are shorthands we will use when we introduce orderings:

Shorthand	Definition
$a \geq 0$	$a > 0$ or $a = 0$
$a > b$	$a - b > 0$
$a < b$	$b > a$

By the way, don't think of a shorthand as something *new*. It's not. It's *just* another way to write something. Don't be scared of it!

If you ever feel like panicking when you see new notation, remember:

```
Math Mantra:   Don't fear notation!   Just expand the notation according to its
               definition to get something you are familiar with.
```

5.7   Ordering Axioms

Recall that all the previous theorems apply to *any field*. In particular, the previous result hold true for the *rationals*.

In order to axiomize the *reals* and *distinguish* them from the *rationals*, we need to introduce one *key axiom*. However, for this axiom to make any sense, we must first introduce the *ordering axioms*.[1]

---

[1]As a fun exercise, you can show that the complex numbers **do not satisfy** the ordering axioms. Therefore, the ordering axioms *distinguish* the reals from the complex numbers.

ORDERING AXIOMS

- **Trichotomy:** *For any $a \in \mathbb{R}$, **exactly** one of the following must hold:*

$$a > 0$$
$$a = 0$$
$$-a > 0$$

- **Positivity:** *For any $x, y \in \mathbb{R}$, if $x > 0$ and $y > 0$, then*

$$x + y > 0$$
$$x \times y > 0$$

Take note of *trichotomy* in particular: it says that a is the same object as 0, or either it or its additive inverse is "greater[1]" than 0. Moreover, trichotomy prescribes that *precisely* one of these conditions holds. This is useful because if we can ever show that two conditions hold, then **we have a contradiction**.

Here are a few consequences of the ordering axioms:

Theorem.

$$1 > 0$$

Proof: Suppose that it is not the case that $1 > 0$. Then by *trichotomy* either

$$1 = 0$$

or

$$-1 > 0.$$

By definition of a field, $1 \neq 0$ so it must be the case that

$$-1 > 0.$$

Using *positivity*,

$$\underbrace{-1}_{x} \times \underbrace{-1}_{y} > 0.$$

But we proved $-1 \times -1 = 1$, so

$$1 > 0$$

which is a contradiction since we assumed this was not the case. Thus $1 > 0$. $\qquad\square$

The next two theorems order elements relative to their inverses: if a is "greater" than 0, its additive inverse is "less" than 0 while its multiplicative inverse is "greater" than 0.

---

[1]I hesitate to say "greater" here since $>$ is just a symbol with the prescribed properties

Theorem. *If*

$$a > 0$$

then

$$0 > -a$$

Proof: We will prove that the two statements are equivalent. The inequality

$$0 > -a$$

is really just *shorthand* for

$$0 - (-a) > 0.$$

Moreover, $0 - (-a)$ is still shorthand, which we can expand to get:

$$0 + (-(-a)) > 0.$$

We already proved the group property

$$(-(-a)) = a,$$

so the inequality is equivalent to

$$0 + a > 0.$$

Since 0 is the additive identity, this is just

$$a > 0$$

as needed. $\square$

Theorem. *If*

$$a > 0$$

then

$$\frac{1}{a} > 0$$

Proof: Let $a > 0$, suppose it is not the case that $\frac{1}{a} > 0$. Then, by *trichotomy* on $\frac{1}{a}$ we either have

$$\frac{1}{a} = 0$$

or

$$-\frac{1}{a} > 0$$

- CASE 1: $\frac{1}{a} = 0$.
  If this is the case, we multiply both sides by a to get

$$1 = 0$$

which is impossible.

- CASE 2: $-\frac{1}{a} > 0$.

  We know that $a > 0$, so by *positivity* on

$$
\begin{aligned}
x &= -\frac{1}{a} \\
y &= a
\end{aligned}
$$

we have

$$
\underbrace{-\frac{1}{a}}_{x} \times \underbrace{a}_{y} > 0
$$

But the left hand side is just

$$
\begin{aligned}
-\tfrac{1}{a} \times a &= (-1 \times a^{-1}) \times a && (-1 \times x = -x) \\
&= -1 \times (a^{-1} \times a) && \text{(Associativity)} \\
&= -1 \times 1 && \text{(Multiplicative Inverse)} \\
&= -1 && \text{(Identity)}
\end{aligned}
$$

So,

$$
-1 > 0.
$$

But this contradicts *trichotomy* since we already know $1 > 0$ and *exactly* one of the three scenarios must hold. Thus, this case is impossible.

Since each case yields a contradiction, we can conclude

$$
\frac{1}{a} > 0. \qquad \square
$$

5.8 Completeness Axiom

Notice that the field axiom and ordering axioms apply to both the *reals* and the *rationals*. Therefore, we need an additional axiom to *distinguish the two*. Precisely, we need an axiom that implies

Irrational numbers exist.

To discover the missing axiom, first view the rationals on a number line where the "holes" are the missing irrational numbers:

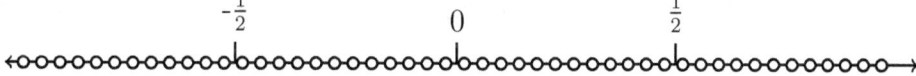

The key idea is to define an axiom that *fills in* these "holes."

Consider some "hole" S on the number line:

This hole can be associated to some set A of elements "left of" S:

Now, we create an axiom that says we can fill in this point, i.e. this point *exists*

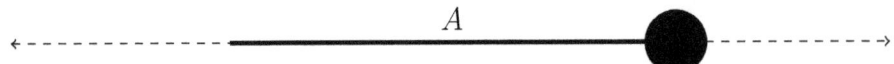

But how can we do this *precisely*?

Let's examine the relationship between the set A and the point S.

First notice that A is to left of S. As per our number line convention, S is greater than any point in set A.

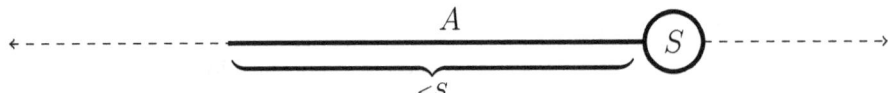

So we say that S is an *upper bound* of set A.

However, A has infinitely many upper bounds, namely all the points to the right of A:

Therefore, in order to *distinguish* S, we notice that S is the *least* of these points to the right of A. Formally,

Definition. *S is the **supremum** or **least upper bound** of the set A if it is both:*

- **An upper bound.**
  S is greater than any element in the set:

  $$x \leq S$$

  for any $x \in A$.

- **The least upper bound.**
  S is smaller than (or equal to) every other upper bound: if there is some B such that for any $x \in A$,

  $$x \leq B,$$

  it is the case that

  $$S \leq B.$$

*Note that the supremum **need not be in S**.*

Armed with this idea, we now define the axiom[1] that *distinguishes* the reals from the rationals:

---

[1]If you are like me, and don't like Hole-ly things, we will have a much nicer equivalent characterization of this axiom once we discuss limits.

Completeness Axiom

> Any non-empty set which is bounded above has a supremum.

Notice that we had to add a few provisos to our axiom, namely the words **bounded above** and **non-empty**. This is because,

- A set that is not bounded above has no rightmost point (the set contains points that are arbitrarily far right).

- An empty set has no rightmost point(what is the rightmost point of nothing)?

The Completeness Axiom gives us a new proof technique: we can now justify certain numbers exist!

5.9 Proof Technique: Existence

I think, therefore I am.

-Descartes

It is easy to prove something doesn't exist. For example, it is easy to prove that we can't have an integer that is both even and odd: if it existed, we would have a contradiction. But,

How do we prove something exists?

The most intuitive answer is

Find It

For example, to prove there exists a number that is divisible by 2 or 3, you can construct 6 as an example. Or, to prove double rainbows exist, you just do a YouTube search.

But besides construction, there are *other ways* to prove existence.

For example, using proof by cases, we already proved that there exist rationals a, b such that

$$a^b$$

is irrational. Contrary to our intuitive technique for proving existence, this proof gives us *no insight* into what such a number looks like. We will see this again next lecture: we will prove

Every non-trivial subspace has a basis

without constructing a particular basis! It's like the author of this book: this book is proof that I exist, but you have no clue what I look like!

The only *real*[1] technique you will use to prove existence is the *Completeness Axiom*. Formally, suppose we want to prove there exists an element that satisfies a specific property. Then,

---

[1]Ignoring my delightful pun, the *Completeness Axiom* is an *axiom*. So proof really means an *agreed acceptance*.

1. Choose some bounded set.

2. By the Completeness Axiom, there exists an element that is a least upper bound of this set.

3. Show that this element satisfies the desired property.

Let's prove the classic example that there is some real number x such that

$$x^2 = 2.$$

To make our lives easier, assume all the typical properties about the rationals and $>$ and do not refer back to the other axioms.[1]

Theorem. *There exists a real number x such that*

$$x^2 = 2.$$

Proof Summary:

- Consider the set of positive numbers whose squares are less than 2.

- The set is bounded and non-empty; therefore, there exists a supremum S by Completeness Axiom.

- $S^2 < 2$ impossible: else you can construct $(S + \epsilon)^2 < 2$, contradicting S is an upper bound.

- $S^2 > 2$ impossible: else you can construct $(S - \epsilon)^2 > 2$, contradicting S is the *least* upper bound.

- Conclude $S^2 = 2$.

Proof: Consider the set

$$\left\{ x \in \mathbb{R} \mid 0 < x \text{ and } x^2 < 2 \right\}$$

First notice this set is non-empty since it contains 1:

$$1^2 = 1 < 2$$

The set is also bounded above by 2. Otherwise, if there was an element y in the set such that $y > 2$, then

$$y^2 > 4$$

But by definition,

$$y^2 < 2$$

so

$$4 < 2$$

---

[1]Before we return to analysis, I would like to make a recommendation. If you enjoyed manipulating abstract properties, I highly recommend taking **Math 120** and **Math 121**. Especially with Professor Sound: he is the *Morgan Freeman* of the Math department!

which is impossible. Thus, by the Completeness Axiom, a supremum S exists. That's the easy part.

Here's the hard part: we have to *check that this element satisfies*

$$S^2 = 2$$

Suppose not. Then either:

$$S^2 < 2 \quad \text{OR} \quad S^2 > 2$$

- CASE 1: $S^2 < 2$.
  To prove this is impossible, we want to find an element that is **bigger than S and in our original set.** This will contradict that S is an *upper bound.* In particular, we need to find a positive ϵ such that

  $$(S + \epsilon)^2 < 2$$

  If we can find such an ϵ, then both

  $$(S + \epsilon) > S$$

  and

  $$(S + \epsilon) \in \left\{ x \in \mathbb{R} \,\middle|\, 0 < x \text{ and } x^2 < 2 \right\}.$$

  But finding such a satisfying ϵ means we must find one that satisfies

  $$\underbrace{S^2 + 2S\epsilon + \epsilon^2}_{(S+\epsilon)^2} < 2$$

  or equivalently,

  $$2S\epsilon + \epsilon^2 < 2 - S^2. \qquad (\star)$$

  Thus, if we can *choose* a positive ϵ that makes the last condition true, we are done!

  But there is a problem: how can we work with an ugly ϵ^2 term? If life were easy and it weren't a square, we could isolate ϵ on one side.

  Here's a trick you are going to use often in this course:

  $$\textbf{If } 0 < \epsilon < 1\textbf{, then } \epsilon^2 < \epsilon$$

  Why is this true? Simple:

  $$\begin{aligned} \epsilon \cdot \epsilon \;&<\; 1 \cdot \epsilon \quad (\text{since } \epsilon < 1) \\ &=\; \epsilon. \end{aligned}$$

  If we restrict $\epsilon < 1$, then we have an upper bound of $2S\epsilon + \epsilon^2$, namely,

  $$2S\epsilon + \epsilon^2 < 2S\epsilon + \epsilon$$

  Therefore, to show $(\star)$ is satisfied, we simply need to show our *upper bound* of $2S\epsilon + \epsilon^2$ is smaller than $2 - S^2$:

  $$2S\epsilon + \epsilon < 2 - S^2$$

But this is a *much easier* problem since we can isolate ϵ

$$\underbrace{\epsilon(2S+1)}_{2S\epsilon+\epsilon} < 2 - S^2,$$

and rewrite it as

$$\epsilon < \frac{2 - S^2}{2S + 1}.$$

Therefore, given the constant S, we need to *find* a positive ϵ that satisfies the above inequality. But that is easy: define ϵ to be the right hand side divided by 2!

$$\epsilon = \frac{2 - S^2}{2(2S + 1)}$$

We are almost good to go, but there is a slight fly in the ointment!

We need to make sure that our ϵ is smaller than 1 for our inequality to work.

Therefore, choose

$$\epsilon = \min\left\{\frac{2 - S^2}{2S + 1}, 1\right\}$$

- CASE 2: $S^2 > 2$
  To prove this is impossible, we find a smaller element that is an upper bound of our set. Particularly, we want to find a positive ϵ such that

$$(S - \epsilon)^2 > 2.$$

Indeed, if this is the case

$$S - \epsilon < S$$

contradicting that S is the *least* upper bound.

By algebra again, this is equivalent to showing

$$-2\epsilon + \epsilon^2 > 2 - S^2.$$

After dividing both sides by -1, this is equivalent to

$$2\epsilon - \epsilon^2 < S^2 - 2.$$

Since $\epsilon^2 > 0$,

$$2\epsilon - \epsilon^2 < 2\epsilon.$$

This means we just have to make sure

$$2\epsilon < S^2 - 2.$$

Simply choose

$$\epsilon = \frac{S^2 - 2}{4}.$$

Lecture 6

All Your Basis are Belong to Us

The best proofs are simple. I did not say easy, I said simple.

-Leon Simon

Goals: In Lecture 3, we learned how to construct a subspace by taking the span of a base set of vectors. Today, we focus on the opposite direction: constructing a base set of spanning vectors *given* the subspace.

6.1 Building Downwards

Whenever we define a new type of animal[1] in mathematics, we like to ask ourselves two major questions. The first is,

Starting from some simple atoms, is there a process to construct animals of that type?

If the answer is yes, then we can build animals to play with. This is the *upwards* approach. The second important question we can ask is,

Starting from the animal, can we break it down into simple atoms?

The reason why we like to ask this question is that

> Math Mantra: If we can break an object into a simple atomic structure, then we can derive tons of properties of the object just by examining the little atoms!

We've seen these upwards and downwards processes before: starting from the base set of prime numbers, we can build all positive integers, and starting from the positive integers, we can decompose them into a product of primes. By studying the prime factorization, we can prove a ridiculous number of cool properties about the integers.

Don't believe that these are important processes? Here is an undergraduate check-list:

---

[1]Another Leon Simonism

115

Area of Mathematics	Topic
Number Theory	Prime Factorization
Algebra	Group Generators
Real Analysis	Open Sets into Intervals
Analysis	Vector Space[1] as a Basis Span
Geometry	Simple Polygons into Triangles
Logic	Syntax
Discrete Math	Tree Growing Procedures

In Lecture 3, we built a subspace by taking the span of a set of vectors. In fact, by removing "redundancy," we could have assumed the span is built from a linearly independent set. That was going *upwards*.

Today, we are going *downwards*. Given the subspace, we will construct a set of spanning vectors. This theorem is literally one of the most important theorems of your undergraduate career and will be used in **many** of your Linear Algebra proofs:

Every non-trivial[2] subspace has a basis.

But before this proof can make sense, you need to learn how to prove that two sets are equal. This vital technique will be used in **all** of your future math courses.

6.2   Proof Technique: Proving Two Sets are Equal

What does it mean for two objects to be equal? In your mathematical career, you will see **many** different notions of equality. Up until this point, the definition has been

Two objects are equal if they are identically the same object.

For example, when I taught in Taiwan, my boss Burch worked as a nightclub DJ. His name was DJ Kimchi, so

Burch = DJ Kimchi

since both names *reference* the same object. So if I hit Burch in the head, that would be the same as hitting DJ Kimchi in the head.

As a more mathematical example, suppose y is a constant satisfying

$$e^y = 2$$

When you applied the log operation on "both sides," what you were really doing was asserting:

Because e^y and 2 are the same object, the log of e^y is the same as the log of 2.

---

[1]Assuming the vector space is finite dimensional. In Math 171, you will see that this is not true for infinite dimensional vector spaces.

[2]The trivial case $V = \{\vec{0}\}$ does not have a basis.

But what do we mean by *two sets are equal*? Intuitively, we know

$$\{a, b, c\}$$

is considered the same as

$$\{c, a, b\}$$

because they have the *same* elements.[1] Formally, we define the following axiom:

If $A \subseteq B$ and $B \subseteq A$, then $A = B$.

In words, this says

If we can show that every element in A is an element of B, and that every element in B is an element of A, then we conclude A and B are the same set.

The logicians call this the *Axiom of Extensionality*.

Let's do a few examples:

Example. *The set of all elements formed from adding integer combinations of 12 and 14 is the same as the set of all even integers:*

$$\{12x + 14y \mid x, y \in \mathbb{Z}\} = \{2n \mid n \in \mathbb{Z}\}$$

Proof: Define
$$\begin{aligned} A &= \{12x + 14y \mid x, y \in \mathbb{Z}\} \\ B &= \{2n \mid n \in \mathbb{Z}\} \end{aligned}$$

- $\subseteq$

  Let a be an arbitrary element of A. By definition of A,

  $$a = 12s + 14t$$

  for some integers s, t. Then,

  $$a = 12s + 14t = 2 \underbrace{(6s + 7t)}_{\in \mathbb{Z}}.$$

  Thus, $a \in B$. Since a was arbitrary, every element of A is in B.

- $\supseteq$

  Now let b be an arbitrary element of B. Then,

  $$b = 2j$$

  for some integer j. But

  $$b = 2j = (14 - 12)j = 12(-j) + 14j$$

  Therefore, $b \in A$. Since b was arbitrary, every element of B is in A.

---

[1]However, $(a, b, c) \neq (c, a, b)$. Ordered tuples are not the same as sets.

Since $A \subseteq B$ and $B \subseteq A$, we can conclude $A = B$. $\qquad\square$

In **Lecture 4**, we mentioned that applying elementary operations to a system of equations leaves the solution space unchanged. This is equivalent to showing

The **set of solutions of the original system** is the same as
the **set of solutions of the transformed system**.

Since this is trivial to show for switching two equations and scaling one equation by a non-zero constant, we only prove the last transformation:

Example. *Consider the system of equations*

$$
\begin{array}{ccccccccccc}
a_{11}x_1 & + & a_{12}x_2 & + & a_{13}x_3 & + & \ldots & + & a_{1n}x_n & = & b_1 \\
a_{21}x_1 & + & a_{22}x_2 & + & a_{23}x_3 & + & \ldots & + & a_{2n}x_n & = & b_2 \\
& \vdots & & \vdots & & \vdots & & \vdots & & \vdots & \\
a_{m1}x_1 & + & a_{m2}x_2 & + & a_{m3}x_3 & + & \ldots & + & a_{mn}x_n & = & b_m
\end{array}
$$

and the system obtained by taking the first equation and adding a scalar k times another equation:

$$
\begin{array}{ccccccccccc}
(a_{11}+ka_{i1})x_1 & + & (a_{12}+ka_{i2})x_2 & + & (a_{13}+ka_{i3})x_3 & + & \ldots & + & (a_{1n}+ka_{in})x_n & = & b_1+kb_i \\
a_{21}x_1 & + & & & a_{22}x_2 & + & a_{23}x_3 & + \ldots + & a_{2n}x_n & = & b_2 \\
& \vdots & & & & \vdots & & \vdots \quad \vdots & & \vdots & \\
a_{m1}x_1 & + & & & a_{m2}x_2 & + & a_{m3}x_3 & + \ldots + & a_{mn}x_n & = & b_m
\end{array}
$$

Then

$$S_1 = S_2$$

where S_1 is the set of all solutions to the first system and S_2 is the set of all solutions to the transformed system.

Proof:

- $\subseteq$

  Let $s \in S_1$,

$$s = (s_1, s_2, \ldots, s_n)$$

  By definition,

$$
\begin{array}{ccccccccccc}
a_{21}s_1 & + & a_{22}s_2 & + & a_{23}s_3 & + & \ldots & + & a_{2n}s_n & = & b_2 \\
& \vdots & & \vdots & & \vdots & & \vdots & & \vdots & \\
a_{m1}s_1 & + & a_{m2}s_2 & + & a_{m3}s_3 & + & \ldots & + & a_{mn}s_n & = & b_m
\end{array}
$$

  So we just need to check that

$$(a_{11}+ka_{i1})s_1 + (a_{12}+ka_{i2})s_2 + (a_{13}+ka_{i3})s_3 + \ldots + (a_{1n}+ka_{in})s_n = b_1 + kb_i$$

  But

$$a_{i1}s_1 + a_{i2}s_2 + a_{i3}s_3 + \ldots + a_{in}s_n = b_i$$

implies
$$ka_{i1}s_1 + ka_{i2}s_2 + ka_{i3}s_3 + \ldots + ka_{in}s_n = kb_i$$

and we are already given
$$a_{11}s_1 + a_{12}s_2 + a_{13}s_3 + \ldots + a_{1n}s_n = b_1.$$

Summing the two equations yields
$$(a_{11} + ka_{i1})s_1 + (a_{12} + ka_{i2})s_2 + (a_{13} + ka_{i3})s_3 + \ldots + (a_{1n} + ka_{in})s_n = b_1 + kb_i.$$

Thus, $s \in S_2$

- $\supseteq$

Now assume $s \in S_2$. As before, s satisfies equations $2, 3 \ldots, m$ so we need only show that
$$a_{11}s_1 + a_{12}s_2 + a_{13}s_3 + \ldots + a_{1n}s_n = b_1$$

But we already know
$$a_{i1}s_1 + a_{i2}s_2 + a_{i3}s_3 + \ldots + a_{in}s_n = b_i.$$

so scaling both sides by k,
$$ka_{i1}s_1 + ka_{i2}s_2 + ka_{i3}s_3 + \ldots + ka_{in}s_n = kb_i.$$

Subtracting this from
$$(a_{11} + ka_{i1})s_1 + (a_{12} + ka_{i2})s_2 + (a_{13} + ka_{i3})s_3 + \ldots + (a_{1n} + ka_{in})s_n = b_1 + kb_i$$

yields
$$a_{11}s_1 + a_{12}s_2 + a_{13}s_3 + \ldots + a_{1n}s_n = b_1$$

so $s \in S_1$. □

In Lecture 5, we derived the division algorithm to get expressions of the form:
$$n = qb + r.$$

We can exploit this algorithm to rapidly calculate the greatest common divisor between two integers. This calculation relies on the following fact:

Example. *For any integers n, b, q, r such that*

$$n = qb + r$$

we have

$$\gcd(n, b) = \gcd(b, r)$$

Proof: Define
$$A = \{d \in \mathbb{Z} \mid d \text{ divides } n \text{ and } d \text{ divides } b\}$$
$$B = \{d \in \mathbb{Z} \mid d \text{ divides } b \text{ and } d \text{ divides } r\}$$

i.e

A is the set of common divisors of n and b

B is the set of common divisors of b and r

If we can prove A and B are the same set, then, in particular, the greatest element in each set is the same. Thus,

$$\gcd(n, b) = \gcd(b, r).$$

- $\subseteq$

  Let a be an arbitrary element of A. Then a divides n and b, so

$$
\begin{aligned}
n &= t_1 a \\
b &= t_2 a
\end{aligned}
$$

for some integers t_1, t_2. Plugging in

$$\underbrace{t_1 a}_{n} = q \underbrace{t_2 a}_{b} + r$$

we get

$$r = (t_1 - q t_2)a$$

so a divides r. Since we already know a divides b, we conclude $a \in B$. Since a was arbitrary, every element of A is in B.

- $\supseteq$

  Follow the same argument: let s be an arbitrary element of B. Then,

$$
\begin{aligned}
b &= t_1 s \\
r &= t_2 s
\end{aligned}
$$

for some integers t_1, t_2. Again, plug in:

$$n = q \underbrace{t_1 s}_{b} + \underbrace{t_2 s}_{r}$$

Then,

$$n = (q t_1 + t_2)s$$

so s divides n. Thus, $s \in A$. Since s was an arbitrary element of B, every element of B is in A.

In conclusion,

$$\{d \in \mathbb{Z}\mid d \text{ divides } n \text{ and } d \text{ divides } b\} = \{d \in \mathbb{Z}\mid d \text{ divides } b \text{ and } d \text{ divides } r\}$$

Particularly, the greatest element in each set is the same, implying:

$$\gcd(n, b) = \gcd(b, r). \qquad \square$$

Why is the preceding theorem useful? Suppose you wanted to compute

$$\gcd(179217921, 17921792)$$

The elementary school way to compute this is to calculate the lists of factors and take the biggest number on both lists:

$$17921792 \;:\; 1,\; 2,\quad 4 \;\ldots$$
$$179217921 \;:\; 1,\; 3,\quad 373 \;\ldots$$

This is painfully slow. Instead, we can use the theorem we just proved.

$$\underbrace{179217921}_{n} = \underbrace{10}_{q} \cdot \underbrace{17921792}_{b} + \underbrace{1}_{r}$$

Immediately,

$$\gcd(179217921, 17921792) = \gcd(17921792, 1) = 1$$

Number theory is great, but how about a more relevant example? Absolutely!

Here's a simple one: consider the vector with i-th component 1 and the rest 0:

$$\vec{e}_i = \begin{bmatrix} 0 \\ \vdots \\ 0 \\ 1 \\ 0 \\ \vdots \\ 0 \end{bmatrix} \longleftarrow i\text{-th component}$$

This vector $\vec{e}_i$ is called the **i-th standard basis vector**. Why are the standard basis vectors $\vec{e}_1, \vec{e}_2, \ldots, \vec{e}_n$ important? We can break any vector in $\mathbb{R}^n$ into a sum of scaled standard basis vectors.

For example,

$$\begin{bmatrix} 3 \\ 2 \\ 1 \end{bmatrix} = 3 \begin{bmatrix} 1 \\ 0 \\ 0 \end{bmatrix} + 2 \begin{bmatrix} 0 \\ 1 \\ 0 \end{bmatrix} + 1 \begin{bmatrix} 0 \\ 0 \\ 1 \end{bmatrix} = 3\vec{e}_1 + 2\vec{e}_2 + 1\vec{e}_3$$

Rigorously,

Example. *For standard basis vectors $\vec{e}_1, \vec{e}_2, \ldots, \vec{e}_n \in \mathbb{R}^n$,*

$$\mathbb{R}^n = \operatorname{span}\{\vec{e}_1, \vec{e}_2, \ldots, \vec{e}_n\}$$

Proof:

- $\supseteq$
  This is immediate since any span of vectors in $\mathbb{R}^n$ is contained in $\mathbb{R}^n$.

- $\subseteq$
  Let $\vec{x} \in \mathbb{R}^n$. Then,

$$\vec{x} = \begin{bmatrix} x_1 \\ x_2 \\ \vdots \\ x_n \end{bmatrix} = x_1 \begin{bmatrix} 1 \\ 0 \\ \vdots \\ 0 \end{bmatrix} + x_2 \begin{bmatrix} 0 \\ 1 \\ \vdots \\ 0 \end{bmatrix} + \ldots + x_n \begin{bmatrix} 0 \\ 0 \\ \vdots \\ 1 \end{bmatrix} = x_1 \vec{e}_1 + x_2 \vec{e}_1 + \ldots + x_n \vec{e}_n$$

Thus,

$$\vec{x} \in \text{span}\{\vec{e}_1, \vec{e}_2, \ldots, \vec{e}_n\} \qquad \square$$

The next example is an often-used span property. Specifically, if we have a "redundant vector" in our spanning list, we can toss it away:

Example. *Suppose $\vec{v}$ is a linear combination of $\vec{v}_1, \vec{v}_2, \ldots, \vec{v}_n$. Then,*

$$\text{span}\{\vec{v}_1, \vec{v}_2, \ldots, \vec{v}_n, \vec{v}\} = \text{span}\{\vec{v}_1, \vec{v}_2, \ldots, \vec{v}_n\}$$

Proof:

- $\supseteq$
  This is immediate from definition. Any linear combination of the vectors on the right is a linear combination of the vectors on the left if we just add $0\vec{v}$.

- $\subseteq$
  Let

$$\vec{x} \in \text{span}\{\vec{v}_1, \vec{v}_2, \ldots, \vec{v}_n, \vec{v}\}$$

Then,

$$\vec{x} = \alpha_1 \vec{v}_1 + \alpha_2 \vec{v}_2 + \ldots + \alpha_n \vec{v}_n + \alpha \vec{v}$$

for some $\alpha_1, \ldots, \alpha_n, \alpha \in \mathbb{R}$. But $\vec{v}$ is a linear combination of the other vectors:

$$\vec{v} = \beta_1 \vec{v}_1 + \beta_2 \vec{v}_2 + \ldots + \beta_n \vec{v}_n$$

for some $\beta_1, \ldots, \beta_n \in \mathbb{R}$. Substitute this back into $\vec{x}$:

$$\vec{x} = \alpha_1 \vec{v}_1 + \alpha_2 \vec{v}_2 + \ldots + \alpha_n \vec{v}_n + \alpha \underbrace{(\beta_1 \vec{v}_1 + \beta_2 \vec{v}_2 + \ldots + \beta_n \vec{v}_n)}_{\vec{v}}.$$

Distribute:

$$\vec{x} = \alpha_1 \vec{v}_1 + \alpha_2 \vec{v}_2 + \ldots + \alpha_n \vec{v}_n + \alpha\beta_1 \vec{v}_1 + \alpha\beta_2 \vec{v}_2 + \ldots + \alpha\beta_n \vec{v}_n$$

and group terms:

$$\vec{x} = (\alpha_1 + \alpha\beta_1)\vec{v}_1 + (\alpha_2 + \alpha\beta_2)\vec{v}_2 + \ldots + (\alpha_n + \alpha\beta_n)\vec{v}_n.$$

Thus,

$$\vec{x} \in \text{span}\{\vec{v}_1, \vec{v}_2, \ldots, \vec{v}_n\}. \qquad \square$$

For the next example, let's prove a set theoretic property. Here's one of *De Morgan's Laws*:

Example. *For any sets A and B,*
$$(A \cup B)^c = A^c \cap B^c$$

Proof:

- $\subseteq$

  Let x be an arbitrary element of $(A \cup B)^c$. Then by definition,
  $$x \notin A \cup B \qquad (\star)$$

  Suppose, for a contradiction, $x \in A$. Then, by definition of union,
  $$x \in A \cup B$$

  But this contradicts $(\star)$. Thus, $x \notin A$. In other words,
  $$x \in A^c$$

  Likewise, suppose $x \in B$. Then, by definition of union,
  $$x \in A \cup B$$

  Again, this contradicts $(\star)$, allowing us to conclude
  $$x \in B^c.$$

  By definition of intersection
  $$x \in A^c \cap B^c$$

- $\supseteq$

  Let x be an arbitrary element of $A^c \cap B^c$. By definition,
  $$\begin{aligned} x &\notin A \\ x &\notin B \end{aligned}$$

  Suppose, for an eventual contradiction,
  $$x \notin (A \cup B)^c$$

  This means
  $$x \in A \cup B.$$

  By definition of *union*, we know either
  $$x \in A$$
  or
  $$x \in B.$$

  In either case, we have a contradiction since we already know $x \notin A$ and $x \notin B$. In conclusion,
  $$x \in (A \cup B)^c. \qquad \square$$

6.3   The Basis Theorem: Showing a Basis Exists

First we define

Definition. *A **basis** for a subspace $V \subseteq \mathbb{R}^n$ is a **finite** set of vectors*

$$\vec{v}_1, \vec{v}_2, \ldots, \vec{v}_n$$

such that:

- $\{\vec{v}_1, \vec{v}_2, \ldots, \vec{v}_n\}$ *is linearly independent.*

- $\text{span}\{\vec{v}_1, \vec{v}_2, \ldots, \vec{v}_n\} = V.$

We are going to show that every non-trivial subspace has a basis. But how do we begin?

Existence proofs are often very difficult, mostly because you don't even know where to start. To quote Professor Simon,

You just need a nice, simple, down to Earth idea.

Here's the idea: let's just consider **every** combination of vectors in V. The biggest linearly independent set that exists is going to be the basis!

Simple right? There are, however, a few questions we need to ask:

1. **How do we know that there is a biggest linearly independent set?** Maybe I can keep on finding arbitrarily large linearly independent sets:

$$\{\vec{a}_1, \vec{a}_2, \vec{a}_3\}$$
$$\{\vec{b}_1, \vec{b}_2, \vec{b}_3, \vec{b}_4, \vec{b}_5, \vec{b}_6, \vec{b}_7, \vec{b}_8\}$$
$$\{\vec{c}_1, \vec{c}_2, \vec{c}_3, \vec{c}_4, \vec{c}_5, \vec{c}_6, \vec{c}_7, \vec{c}_8, \vec{c}_9, \vec{c}_{10}, \vec{c}_{11}, \vec{c}_{12}, \vec{c}_{13}, \ldots, \vec{c}_{1792}\}$$
$$\vdots$$

2. **Suppose a biggest linearly independent set does exist. Are we *allowed* to consider every possible combination of vectors in V and *choose* a biggest linearly independent set from this collection?** We can't physically test every combination of vectors for linear independence or even write them all out!

The first question is easy: by the Linear Dependence Lemma, we are going to show that we cannot have more than n linearly independent vectors in $\mathbb{R}^n$.

The second question is far more difficult because it is *philosophical.*

Considering all possible combinations of vectors in V and being able to *choose* a maximal linearly independent set is *sketchy.* Very sketchy indeed. It has to do with the thin grey line between working mathematicians and the logicians called the **Axiom of Choice.** We save this for the end of the lecture, but for this course, you must *choose* to accept that we can find a biggest linearly independent set (pun intended).

Theorem (Basis Theorem). *Every nontrivial subspace $V \subseteq \mathbb{R}^n$ has a basis.*

Proof Summary:

- By the Linear Dependence Lemma, V cannot contain more than n linearly independent vectors.

- Since we cannot have arbitrarily large linearly independent sets, the maximum size M must be achieved by some linearly independent set.

- We can show this set is a basis for V.

  - *Linear Independence:* Obvious.
  - *Spanning:*
    * $\subseteq$: Obvious.
    * $\supseteq$: Suppose not. Then $M + 1$ linearly independent vectors in V. But we assumed the biggest set size is M, contradiction.

Proof: Let $V \subseteq \mathbb{R}^n$. I claim that V cannot have more than n linearly independent vectors. Suppose it does. Then there exists vectors

$$\vec{v}_1, \vec{v}_2, \ldots, \vec{v}_n, \vec{v}_{n+1}, \vec{v}_{n+2}, \ldots, \vec{v}_{n+k}$$

that are linearly independent. Of course, this implies the first $n + 1$ vectors

$$\vec{v}_1, \vec{v}_2, \ldots, \vec{v}_{n+1}$$

are linearly independent (quick check using contradiction)!

Recall from our examples, we showed

$$\mathbb{R}^n = \text{span}\{\vec{e}_1, \vec{e}_2, \ldots, \vec{e}_n\}.$$

Of course, each of our $\vec{v}$'s are in $\mathbb{R}^n$. But this means we just found $n + 1$ independent vectors in

$$\text{span}\{\vec{e}_1, \vec{e}_2, \ldots, \vec{e}_n\}.$$

This **directly contradicts** Linear Dependence Lemma!

Therefore, we can find at most n linearly independent vectors in V.

Consider all possible combinations of vectors in V. Since we just proved that we cannot have arbitrarily large linearly independent sets, there must be some linearly independent set that achieves the **maximum** size M, with $1 \leq M \leq n$:

$$\vec{v}_1, \vec{v}_2, \vec{v}_3, \ldots, \vec{v}_M.$$

Note, I did not tell you how to find this set of vectors. We know it exists[1] by considering **all possible** vector combinations.

I claim this set is a basis. Since it is linearly independent by definition, we just need to show

$$\text{span}\{\vec{v}_1, \vec{v}_2, \ldots, \vec{v}_M\} = V$$

- $\subseteq$
  Immediate: if
  $$\vec{x} \in \text{span}\{\vec{v}_1, \vec{v}_2, \ldots, \vec{v}_M\}$$
  then
  $$\vec{x} = \alpha_1 \vec{v}_1 + \alpha_2 \vec{v}_2 + \ldots + \alpha_M \vec{v}_M$$
  But each of the $\vec{v}$'s are in V, and V is closed under addition and scaling. So $\vec{x} \in V$.

- $\supseteq$
  Let $\vec{v} \in V$. Suppose
  $$\vec{v} \notin \text{span}\{\vec{v}_1, \vec{v}_2, \ldots, \vec{v}_M\}.$$
  This means
  $$\vec{v}_1, \vec{v}_2, \ldots \vec{v}_M, \vec{v}$$
  is linearly independent (because $\vec{v}$ cannot be written as a linear combination of the other vectors). But this is impossible since we found $M + 1$ linearly independent vectors and we assumed we can have at most M!

Thus,
$$\vec{v} \in \text{span}\{\vec{v}_1, \vec{v}_2, \ldots, \vec{v}_M\}.$$

In conclusion,
$$V = \text{span}\{\vec{v}_1, \vec{v}_2, \ldots, \vec{v}_M\}$$

and since V was an arbitrary subspace, every subspace has a basis. $\square$

6.4   The Basis Theorem, Part II: Finding a Basis

So we showed a basis exists. Cool. But that still doesn't answer the question of how to find it. What if you pick a few initial linearly independent vectors and then hit a dead-end[2]? In this case, it's impossible to add more vectors to make the span V:

---

[1]Remember when Rose tossed the Heart of the Ocean off the Titanic? I have no clue how to find her necklace. I only know it's somewhere in the ocean.

[2]In computer science, you will see dilemmas like this in the "packing problem."

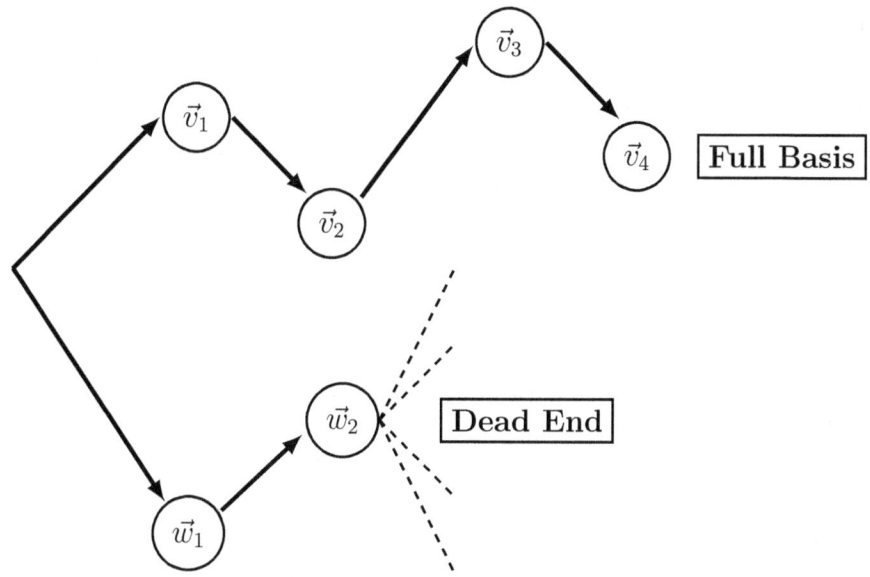

But we can prove this doesn't happen and in fact, it **doesn't matter** what you choose as the first few linearly independent vectors. You can always **extend** them to a full basis. So that means that to find a basis, we just need to keep throwing in any linear independently vectors we can find, including the kitchen sink.[1]

By modifying the Basis Theorem, we can come up with a powerful *extension theorem*:

Theorem (Basis Extension Theorem). *Given linearly independent vectors*

$$\vec{w}_1, \vec{w}_2, \ldots, \vec{w}_k$$

in V, we can always extend this set to a full basis. In other words, we can find $\vec{v}_{k+1}, \vec{v}_{k+2}, \ldots, \vec{v}_M$ such that

$$\vec{w}_1, \vec{w}_2, \ldots, \vec{w}_k, \vec{v}_{k+1}, \vec{v}_{k+2}, \ldots, \vec{v}_M$$

are linearly independent and

$$V = \text{span}\{\vec{w}_1, \vec{w}_2, \ldots, \vec{w}_k, \vec{v}_{k+1}, \vec{v}_{k+2}, \ldots, \vec{v}_M\}$$

Proof Summary:

- V cannot contain more than n linearly independent vectors by Linear Dependence Lemma.

- The maximum size linearly independent set containing original vectors is achieved by some set.

- This set is a basis for V.

Proof: Once again, by Linear Dependence Lemma, we know that we cannot add an unlimited number of linearly independent vectors to the set

$$\vec{w}_1, \vec{w}_2, \ldots, \vec{w}_k.$$

---

[1]Another Leon Simonism.

Otherwise, we will have $n + 1$ linearly independent vectors in $\mathbb{R}^n$.

In the very best case scenario, we can add vectors so our set has maximally M linearly independent vectors

$$\vec{w}_1, \vec{w}_2, \ldots, \vec{w}_k, \vec{v}_{k+1}, \vec{v}_{k+2}, \ldots, \vec{v}_M$$

Again, we show

$$\text{span}\{\vec{w}_1, \vec{w}_2, \ldots, \vec{w}_k, \vec{v}_{k+1}, \vec{v}_{k+2}, \ldots, \vec{v}_M\} = V.$$

- $\subseteq$

  This follows from closure of a subspace under addition and scaling.

- $\supseteq$

  Let $\vec{v} \in V$. Suppose

  $$\vec{v} \notin \text{span}\{\vec{w}_1, \vec{w}_2, \ldots, \vec{w}_k, \vec{v}_{k+1}, \vec{v}_{k+2}, \ldots, \vec{v}_M\}$$

  This implies

  $$\vec{w}_1, \vec{w}_2, \ldots, \vec{w}_k, \vec{v}_{k+1}, \vec{v}_{k+2}, \ldots, \vec{v}_M, \vec{v}$$

  is a linearly independent set. But an $M + 1$ linearly independent set contradicts that we can have at most M linearly independent vectors in V. Thus,

  $$\vec{v} \in \text{span}\{\vec{w}_1, \vec{w}_2, \ldots, \vec{w}_k, \vec{v}_{k+1}, \vec{v}_{k+2}, \ldots, \vec{v}_M\} \qquad \square$$

Therefore, we will *eventually* get a basis by repeatedly adding whatever linearly independent vectors we can find.

6.5   Dimension

In our proof of The Basis Theorem, we found a basis by looking at a *maximal* linearly independent set. Perhaps this was overkill. Specifically, we need to ask ourselves

Is it possible to find a smaller basis?

The answer to this question is **no**:

Theorem. *Every basis for a given subspace V has the same number of vectors.*

Proof Summary:

- Suppose there exists two bases of different sizes.

- The vectors in the bigger basis are linearly independent in the span of the smaller basis.

- This contradicts the Linear Dependence Lemma.

Proof: Suppose not. Then we can find two bases of different sizes,

$$\vec{v}_1, \vec{v}_2, \ldots, \vec{v}_m$$

$$\vec{w}_1, \vec{w}_2, \vec{w}_3, \vec{w}_4, \ldots, \vec{w}_M$$

Without loss of generality, assume the second set is bigger: $m < M$. Choose the first $m + 1$ vectors in the second list

$$\vec{w}_1, \vec{w}_2, \ldots, \vec{w}_{m+1}$$

These are linearly independent in V; thus, we have $m + 1$ linearly independent vectors in

$$\text{span}\{\vec{v}_1, \vec{v}_2, \ldots, \vec{v}_m\}$$

KaBoom! This is a contradiction by the Linear Dependence Lemma. Thus, every basis for a given subspace V has the same number of vectors. □

Notice that this theorem allows us to associate a *unique* number to each subspace; namely, the number of vectors in *every* basis for that subspace.

Definition. *The **dimension** of a subspace V is the number of vectors in a basis that spans V. We denote this by*

$$\dim V$$

Out of convention, we define the dimension of $\{\vec{0}\}$ (the subspace containing only the zero vector) to be 0.

Why should we care about this number? Intuitively,

Dimension tells us how "big" a subspace is.

For example, consider any subspace V of $\mathbb{R}^3$. If $\dim(V) = 0$, we have a single point at the origin:

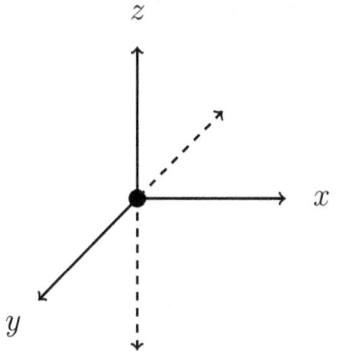

If $\dim(V) = 1$, we have a line:

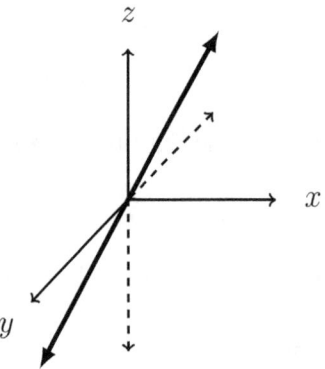

If $\dim(V) = 2$, then we have a plane passing through the origin:

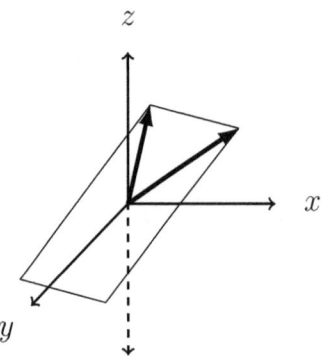

If $\dim(V) = 3$, we have the entire space $\mathbb{R}^3$.

Another reason why dimension is useful is that we can actually use this number to help **find** a basis. Wait,

> *Doesn't dimension come from knowing a particular basis and measuring its size?*

Yes, but you are going to learn how to calculate the dimension V (in some special cases) **without** actually finding a basis.[1] Then, with the following theorem, we can use the dimension to help **find** a basis:

---

[1] All hail the almighty rank nullity theorem!

Theorem. *Let $V \neq \{\vec{0}\}$ be a subspace. Then*

1. *If $\vec{v}_1, \vec{v}_2, \ldots, \vec{v}_{\dim V}$ span V, then they are linearly independent (and thus form a basis for V)*

2. *If $\vec{v}_1, \vec{v}_2, \ldots, \vec{v}_{\dim V}$ are linearly independent, then they span V (and thus form a basis for V)*

Proof Summary:

- 1: Suppose not. Shrink the set of $\vec{v}_i$'s to a set of $\dim k - 1$ elements that still spans V. Then any basis for V is a set of $\dim V$ linearly independent vectors in a span of $\dim k - 1$ vectors, which contradicts Linear Dependence Lemma.

- 2: Suppose not. Extend the set of $\vec{v}_i$'s to a set that spans V and is still linearly independent. This contradicts the fact that any basis has size $\dim V$.

Proof:

- **1.**
  Let
  $$\vec{v}_1, \vec{v}_2, \ldots, \vec{v}_{\dim V}$$
  span V. Suppose they are not linearly independent.

  Recall that, in our examples, we proved that if $\vec{v}$ is a linear combination of $\vec{v}_1, \vec{v}_2, \ldots, \vec{v}_k$, then
  $$\operatorname{span}\{\vec{v}_1, \vec{v}_2, \ldots, \vec{v}_k, \vec{v}\} = \operatorname{span}\{\vec{v}_1, \vec{v}_2, \ldots, \vec{v}_k\}$$

  By linear dependence, at least one of our $\vec{v}_i$ is a linear combination of the others. After possibly reordering the vectors, we may assume that $\vec{v}_1$ is a linear combination of the others. Then,
  $$\operatorname{span}\{\vec{v}_2, \ldots, \vec{v}_{\dim V}\} = \operatorname{span}\{\vec{v}_1, \vec{v}_2, \ldots, \vec{v}_{\dim V}\} = V.$$

  Now select a basis
  $$\vec{w}_1, \vec{w}_2, \ldots, \vec{w}_{\dim V}$$
  for V by the basis theorem. Then, $\vec{w}_1, \vec{w}_2, \ldots, \vec{w}_{\dim V}$ are $\dim V$ linearly independent vectors in $\operatorname{span}\{\vec{v}_2, \ldots, \vec{v}_{\dim V}\}$, contrary to the Linear Dependence Lemma.

- **2.**
  Let
  $$\vec{v}_1, \vec{v}_2, \ldots, \vec{v}_{\dim V}$$
  be linearly independent and suppose this set does not span V. This means that there is some vector $\vec{v} \in V$ that cannot be written as a linear combination of $\vec{v}$'s. Therefore,
  $$\vec{v}_1, \vec{v}_2, \ldots, \vec{v}_{\dim V}, \vec{v}$$
  is linearly independent. Applying our basis extension theorem to this set, a basis would have **strictly more** than $\dim V$ vectors, contradicting uniqueness of basis size! $\qquad\square$

Therefore, if we know the dimension of V, to check that a set of size $\dim V$ is a basis, we only need to check that either the set is linearly independent or the set spans V.

6.6   Some Fun with the Basis Theorems

In Lecture 3, we proved that the intersection of two subspaces is still a subspace. Intuitively, if the intersection is smaller, the basis should be smaller as well. In fact, the new basis should be smaller than the bases of *both* the original subspaces:

Example. *For subspaces $V, W \subseteq \mathbb{R}^n$,*

$$\dim V \cap U \leq \min\{\dim V, \dim U\}$$

Proof: By the basis theorem, we know that $V \cap U$ has a basis

$$\vec{w}_1, \vec{w}_2, \ldots, \vec{w}_k$$

But $V \cap U \subseteq V$, so these $\vec{w}_i$ are linearly independent in V. Therefore, we can extend this to a full basis for V. Thus,

$$\dim V \cap U \leq \dim V$$

Likewise, $V \cap U \subseteq U$, so these $\vec{w}_i$ are linearly independent in U. Therefore, we can extend this to a full basis for U. Thus,

$$\dim V \cap U \leq \dim U.$$

Because the dimension of $V \cap U$ is bounded by *both* $\dim U$ and $\dim V$, it is certainly bounded by the *smaller* of the two:

$$\dim V \cap U \leq \min\{\dim V, \dim U\}. \qquad \square$$

By the way, a **noob mistake** would be to assume a similar result holds for set unions:

$$\dim U \cup V \geq \max\{\dim U, \dim V\}$$

$U \cup V$ need not be a subspace, so taking the dimension does not make sense! $U \cup V$ is an **entirely different animal!** In general,

> Math Mantra: Just because you can write an expression down, doesn't mean it automatically makes sense! You must check that you have the correct animals!

Recall that we also showed that any linear mapping of a subspace is still a subspace. In fact, the image does not "gain" dimension:

Example. *Let V be a subspace of $\mathbb{R}^n$ and let f be a linear map on V i.e.*

$$\begin{aligned} f(\vec{x} + \vec{y}) &= f(\vec{x}) + f(\vec{y}) \quad \text{for all } \vec{x}, \vec{y} \in V \\ f(\alpha\vec{x}) &= \alpha f(\vec{x}) \qquad\quad \text{for all } \vec{x} \in V, \alpha \in \mathbb{R} \end{aligned}$$

Then the image

$$f(V) = \{f(\vec{x}) \mid \vec{x} \in V\}$$

has dimension smaller than or equal to V's dimension:

$$\dim f(V) \leq \dim(V).$$

Proof: Let

$$\vec{v}_1, \vec{v}_2, \ldots, \vec{v}_k$$

be a basis for V (by the Basis Theorem). As a simple consequence of linearity, we can show

$$f(V) = \mathrm{span}\{f(\vec{v}_1), f(\vec{v}_2), \ldots, f(\vec{v}_k)\}$$

- $\subseteq$

  Let $\vec{x} \in f(V)$. By definition,

  $$\vec{x} = f(\vec{v})$$

  for some $\vec{v} \in V$. Since the $\vec{v}_i$'s are a basis for V, we can write

  $$\vec{v} = \alpha_1 \vec{v}_1 + \alpha_2 \vec{v}_2 + \ldots + \alpha_k \vec{v}_k.$$

  for some $\alpha_1, \alpha_2, \ldots, \alpha_k \in \mathbb{R}$. Apply f to both sides:

  $$\underbrace{f(\vec{v})}_{\vec{x}} = f(\alpha_1 \vec{v}_1 + \alpha_2 \vec{v}_2 + \ldots + \alpha_k \vec{v}_k).$$

By linearity,

$$\vec{x} = \alpha_1 f(\vec{v}_1) + \alpha_2 f(\vec{v}_2) + \ldots + \alpha_k f(\vec{v}_k).$$

Thus,

$$\vec{x} \in \mathrm{span}\{f(\vec{v}_1), f(\vec{v}_2), \ldots, f(\vec{v}_k)\}$$

- $\supseteq$

  Let $\vec{x} \in \mathrm{span}\{f(\vec{v}_1), f(\vec{v}_2), \ldots, f(\vec{v}_k)\}$. Then, for some $\alpha_1, \alpha_2, \ldots, \alpha_k \in \mathbb{R}$,

  $$\vec{x} = \alpha_1 f(\vec{v}_1) + \alpha_2 f(\vec{v}_2) + \ldots + \alpha_k f(\vec{v}_k)$$

By linearity, we can combine the f's:

$$\vec{x} = f(\alpha_1 \vec{v}_1 + \alpha_2 \vec{v}_2 + \ldots + \alpha_k \vec{v}_k).$$

Since $\alpha_1 \vec{v}_1 + \alpha_2 \vec{v}_2 + \ldots + \alpha_k \vec{v}_k \in V$,

$$\vec{x} \in f(V)$$

In conclusion,

$$f(V) = \mathrm{span}\{f(\vec{v}_1), f(\vec{v}_2), \ldots, f(\vec{v}_k)\}$$

Therefore, $f(V)$ is a span of k vectors. This means $f(V)$ can have at most k linearly independent vectors in its basis, implying

$$\dim f(V) \leq \underbrace{\dim V}_{k} \qquad \square$$

6.7   Sketchy Shades of Grey: Axiom of Choice

There are many things in the field of Mathematical Logic that the working mathematician will describe as mumbo jumbo. But the *Axiom of Choice* isn't one of them.

If you're not careful, you can assume a seemingly innocuous statement and bad things will happen. For example, if you are an aspiring Logician, instead of axiomatizing mathematics, you may want to look into axiomatizing knowledge and belief.[1] You can take simple axioms like

- *If you believe x, then you believe that you believe x.*

- *If you know x, then you believe x.*

- *If you believe x, then you believe that you know x*

and using simple rules of inference like Modus Ponens, you can derive

Knowing x is equivalent to believing x

But that's wrong:

*I **believe** the Last Airbender is a bad movie*

and

*I **know** the Last Airbender is a bad movie*

are two completely different statements.

What does assuming innocuous axioms have to do with The Basis Theorem? The Basis Theorem implicitly[2] uses *Axiom of Choice*, which asserts

For any collection of sets, we can always find a function on that collection that inputs a non-empty set and outputs an element of that particular set.

We call such a function a *choice* function since it is *choosing* an element from each set. Here's a fun analogy:

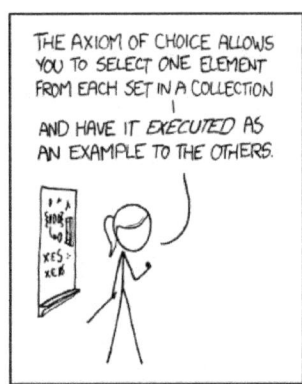

---

[1]Example from *Multi-agent Systems* by Yoav Shoham

[2]In the case of the Basis Theorem, we consider the collection of sets,
{The set of all finite combination of vectors in $V \mid V$ is a non-empty subspace of $\mathbb{R}^n$}. We then construct a function that selects a maximal linearly independent set of vectors from any set in this collection.

For example, consider the sets

$$\{1, 2, 3\} \qquad \{\ldots - 2, -2.00001, 42 \ldots\} \qquad \mathbb{R}^n$$
$$\{0\} \quad \{x | x \in \text{Stanford Math Department}\} \quad \mathbb{R}$$

Constructing a choice function for this collection of sets is easy: go through each set one at a time and choose some element as the output, e.g.

$$
\begin{aligned}
f(\{1, 2, 3\}) &= 2 \\
f(\{\ldots - 2, -2.00001, 42 \ldots\}) &= 42 \\
f(\mathbb{R}^n) &= \vec{e}_2 \\
f(\{0\}) &= 0 \\
f(\{x | x \in \text{Stanford Math Department}\}) &= \text{Soren Galatius} \\
f(\mathbb{R}) &= \pi
\end{aligned}
$$

You can also construct choice functions for infinite collections of sets. For example, consider the collection of all subsets of positive integers. Define f to always spit out the least element in the set:

$$
\begin{aligned}
f(\{2, 4, 6, 8, \ldots\}) &= 2 \\
f(\{72, 83, 94, \ldots\}) &= 72 \\
f(\{101, 1011, 1211, 3333, \ldots\}) &= 101 \\
&\vdots
\end{aligned}
$$

However, suppose I asked you to construct a choice function on the collection of all non-empty subsets of $\mathbb{R}$. It is **humanly impossible** to arbitrarily pick out one element at a time! The problem is that the set is **far too massive**.[1]

In the case of the reals, even though we cannot humanly construct a choice function, it seems like one *should* exist. So we can just take this as axiom, right?

There is an unholy consequence: the Axiom of Choice implies the monster known as the Banach-Tarski Paradox, which states:

We can break a ball into finitely many non-overlapping pieces and rearrange them to form two identical balls:

Weird! For more details on this, I highly recommend reading *The Pea and the Sun* by Leonard Wapner. He gives a nice exposition that requires only a modest amount of mathematics.

Despite these weird consequences, most mathematicians continue to accept the Axiom of Choice.

---

[1]Precisely, the reals are *uncountable*. This will be discussed in the final lecture.

New Notation

Symbol	Reading	Example	Example Translation
$\vec{e}_i$	The i-th standard basis vector	$\text{span}\{\vec{e}_1, \vec{e}_2, \vec{e}_3\} = \mathbb{R}^3$	The span of the first three standard basis vectors is $\mathbb{R}^3$
$A \subseteq B$	A is a subset of or equals B	$\{0\} \subseteq V$	$\{0\}$ is a subset of V or equals V.
$A \supseteq B$	A either equals or is contained in B	$\mathbb{R}^2 \supseteq V$	$\mathbb{R}^2$ contains (or equals) V.
$\dim V$	The dimension of vector space V	$\dim \mathbb{R}^3 = 3$	The dimension of $\mathbb{R}^3$ is 3.

Lecture 7

Matrix Madness

Unfortunately, no one can be told what the Matrix is. You have to practice bookkeeping and working with double sums yourself.

- $\cong$pheus

Goals: Today, we look at matrices and how to rigorously prove matrix properties. We also define a notion of a matrix norm and prove a Cauchy-Schwarz-like inequality. This proof will rely on the key matrix property that the product of a matrix and a vector can be viewed as a linear combination of the matrix columns. Lastly, we prove another key property: any linear function from $\mathbb{R}^m$ to $\mathbb{R}^n$ can be written as a matrix multiplication.

7.1   Let's be Honest

Here's a complete summary of matrices in high school: in Algebra II, you learned how to represent a system of equations as a matrix and row reduce to solve for all the unknowns. Then, you learned how to mindlessly compute matrix products, sums, and inverses without any context as to *why*. A majority of Honors Algebra II teachers stop right there. The really good ones[1] go a little further and say that you can use matrix products to represent the system as a matrix multiplication.

$$A\vec{x} = \vec{b}$$

Then you can compute the inverse of A as long as $\det(A) \neq 0$ and multiply both sides by the inverse to get

$$
\begin{array}{rcll}
A\vec{x} & = & \vec{b} & \Longleftrightarrow \\
A^{-1}A\vec{x} & = & A^{-1}\vec{b} & \Longleftrightarrow \\
\vec{x} & = & A^{-1}\vec{b} &
\end{array}
$$

Typically, you learned this in a supplemental reading from Howard Anton since the standard high school texts are very lacking in Linear Algebra. You also memorized the slogans

Matrix multiplication is associative
Matrix multiplication is not always commutative
For matrix multiplication to work, the inner dimensions must match

---

[1]Shout out to Ms. Evans of Los Altos Hills, Mr. Friedland of Palo Alto High, and Mr. Lazar of San Jose Mission

Now, the soul-shattering question I need to ask you is, *why?*

Why does $\det(A) = 0$ imply non-invertibility?
Why is matrix multiplication always associative?
Why does concatenating a matrix with its identity let you find the inverse?

$$\left[\begin{array}{ccc|ccc} 0 & 2 & 0 & 1 & 0 & 0 \\ 2 & 1 & 0 & 0 & 1 & 0 \\ 3 & 1 & 1 & 0 & 0 & 1 \end{array}\right]$$

But most importantly,

Why should you care?

You could have bypassed matrices and just stuck with systems of equations! It would have been easy since you only dealt with 3×3 and 2×2 matrices.

You also need to ask yourselves the *is* questions:

Is there a way to compute solutions of $Ax = b$ when $\det(A) = 0$?
Is there more use for determinants than checking invertibility?
Is there a greater purpose for matrices then just notation?

Like when Aladdin met Jasmine, the H-series is going to show you a **whole new world** behind matrices.

I used to be in your shoes. I know Gaussian Elimination and computation are a breeze and you will be able to understand the theorem statements (though you may have trouble juggling m and n). There are two things that are going to scare you, mainly because you have never had any practice with them:

- **Working with Σ-notation.**

- **Proving two matrices (of arbitrary size) are equal.**

Let's conquer these fears:

7.2   Working with Sums

It is my experience that Σ-notation is more of a hindrance than a help for beginners in linear algebra. Therefore, I have generally avoided its use.

-Howard Anton

Unfortunately, we won't have this liberty. If you want to survive the H-series, you have to master Σ-notation. True, I will avoid Σ-notation if it makes a concept clearer. However, when you hit *determinants*, this notation will be *completely unavoidable.*

You especially need to be comfortable with this notation when working with matrices. Primarily, Σ-notation condenses *complicated* expressions.

For example, given the matrices

$$A = \begin{bmatrix} a_{11} & a_{12} & \cdots & a_{1n} \\ a_{21} & a_{22} & \cdots & a_{2n} \\ \vdots & \vdots & \ddots & \vdots \\ a_{m1} & a_{m2} & \cdots & a_{mn} \end{bmatrix} \qquad B = \begin{bmatrix} b_{11} & b_{12} & \cdots & b_{1p} \\ b_{21} & b_{22} & \cdots & b_{2p} \\ \vdots & \vdots & \ddots & \vdots \\ b_{n1} & b_{n2} & \cdots & b_{np} \end{bmatrix}$$

the matrix product AB is defined as the matrix

$$AB = \begin{bmatrix} [AB]_{11} & [AB]_{12} & \cdots & [AB]_{1j} & \cdots & [AB]_{1p} \\ \vdots & \vdots & \vdots & \vdots & \vdots & \vdots \\ [AB]_{i1} & [AB]_{i2} & \cdots & [AB]_{ij} & \cdots & [AB]_{ip} \\ \vdots & \vdots & \vdots & \vdots & \vdots & \vdots \\ [AB]_{m1} & [AB]_{m2} & \cdots & [AB]_{mj} & \cdots & [AB]_{mp} \end{bmatrix}$$

where each ij entry $[AB]_{ij}$ is the dot product of the i-th row of A with the j-th column of B:

$$\begin{bmatrix} a_{11} & a_{12} & \cdots & a_{1n} \\ \vdots & \vdots & \vdots & \vdots \\ a_{i1} & a_{i2} & \cdots & a_{in} \\ \vdots & \vdots & \vdots & \vdots \\ a_{m1} & a_{m2} & \cdots & a_{mn} \end{bmatrix} \begin{bmatrix} b_{11} & \cdots & b_{1j} & \cdots & b_{1p} \\ b_{21} & \cdots & b_{2j} & \cdots & b_{2p} \\ \vdots & \vdots & \vdots & \vdots & \vdots \\ b_{n1} & \cdots & b_{nj} & \cdots & b_{np} \end{bmatrix}$$

We can use Σ-notation to condense this definition:

Definition. *Let A be an $m \times n$ matrix and B be an $n \times p$ matrix, then the product AB is defined as the matrix with ij entry*

$$[AB]_{ij} = \sum_{r=1}^{n} a_{ir} b_{rj}.$$

You also need to be comfortable with *double sums*.[1]

Consider the expression

$$\sum_{i=1}^{m} \sum_{j=1}^{n} a_{ij}$$

In actuality, this is a shorthand for one sum *nested* within another. In the inner sum, the i variable is *fixed*:

$$\sum_{i=1}^{m} \left(\sum_{j=1}^{n} a_{ij} \right) = \sum_{i=1}^{m} (a_{i1} + a_{i2} + \ldots a_{in})$$

Of course, we can apply single summation properties to double summations:

---

[1]You've already seen these when proving Cauchy-Schwarz on Homework 1.

Example.

$$\sum_{i=1}^{n}\sum_{j=1}^{n} ij = \left(\sum_{j=1}^{n} j\right)^2$$

Proof: Viewing the left hand side as one sum nested within the other,

$$\sum_{i=1}^{n}\left(\sum_{j=1}^{n} ij\right),$$

we can pull out i from the inner sum since it is a constant (as j varies):

$$\sum_{i=1}^{n} i\left(\sum_{j=1}^{n} j\right).$$

But within the outer sum, notice that

$$\sum_{j=1}^{n} j$$

is a constant (as i varies), so we call pull that out of the outer sum, giving us a product

$$\left(\sum_{j=1}^{n} j\right)\left(\sum_{i=1}^{n} i\right).$$

The dummy variable in each of the summations *doesn't matter*, so we change the i into a j, giving us

$$\left(\sum_{j=1}^{n} j\right)^2 \qquad \square$$

One of the most fundamental properties of double summations is that the Σ's "commute:"

$$\sum_{i=1}^{m}\sum_{j=1}^{n} a_{ij} = \sum_{j=1}^{n}\sum_{i=1}^{m} a_{ij}$$

Visualizing the terms in an array, this equality states that summing over the columns is the same as summing over the rows:

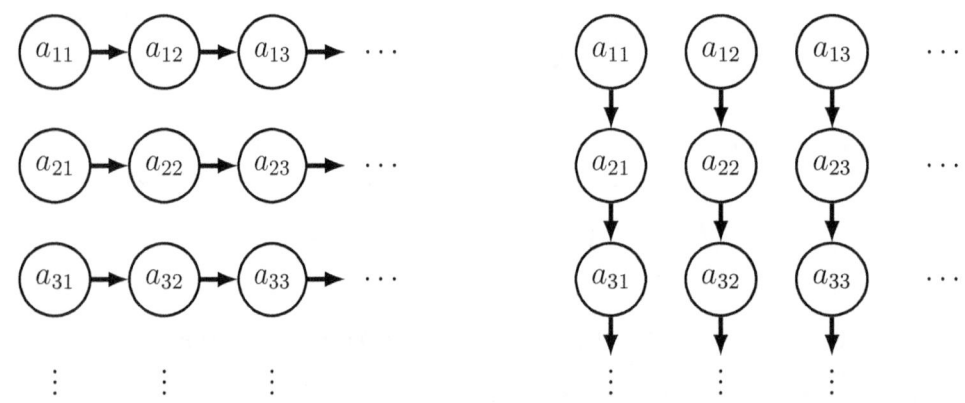

Intuitively, this property is obvious. But how do we prove it rigorously?

If you try to directly apply single summation properties, you may have trouble untangling the i from the j.

Instead of purely algebraic manipulation, we need to look at the *meaning* behind the summation symbol.

> Math Mantra: Instead of jumping to algebraic manipulation, try to understand the INTENT[1] of the notation.

When we write a sum, we are actually summing over a *set* of distinct terms. For example,

$$\sum_{i=1}^{n} a_i$$

is really just shorthand for

Sum all terms of the form a_i where $i \in \{1, 2, 3, \ldots, n\}$

Alternatively, we could have represented this meaning by a different notation:

$$\sum_{i \in \{1,2,3,\ldots,n\}} a_i$$

where we sum over distinct terms $\{1, 2, 3, \ldots, n\}$. The jargon for such a set is an *indexing set*.

Until now, you only worked with a notation that only permitted summations over terms indexed by consecutive integers. But we can write more interesting expressions like

$$\sum_{\{d \mid d > 0,\, d \text{ divides } 24\}} d$$

or

$$\sum_{\{p \leq 15,\, p \text{ is prime}\}} p$$

or even over infinite indexing sets like

$$\sum_{x \in \mathbb{R}} a_x$$

The first two sums, respectively, translate to

$$
\begin{aligned}
1 + 2 + 3 + 4 + 6 + 12 + 24 &= 48 \\
2 + 3 + 5 + 7 + 11 + 13 &= 41
\end{aligned}
$$

However, we must be careful about our notation. The last expression has a meaningless indexing set in the context[2] of sums.

---

[1] You will see this again when proving $\det A = \det A^T$

[2] This indexing set is meaningful, however, in the context of *unions*.

Example.

$$\sum_{i=1}^{m}\sum_{j=1}^{n} a_{ij} = \sum_{j=1}^{n}\sum_{i=1}^{m} a_{ij}$$

Proof: The left hand side is shorthand for

$$\sum_{(i,j)\in S} a_{ij}$$

where

$$S = \{(i,j)|\ i,j \text{ are integers and for each } j \text{ in } \{1,1,2,\ldots m\}, i \text{ is in } \{1,2,\ldots,n\}\}.$$

Likewise,

$$\sum_{j=1}^{n}\sum_{i=1}^{m} a_{ij}$$

is just shorthand for

$$\sum_{(i,j)\in S'} a_{ij}$$

where

$$S' = \{(i,j)|\ i,j \text{ are integers and for each } i \text{ in } \{1,2,\ldots n\}, j \text{ is in } \{1,2,\ldots,m\}\}.$$

But it is a simple exercise to show that $S' = S$, so

$$\sum_{i=1}^{m}\sum_{j=1}^{n} a_{ij} = \sum_{(i,j)\in S} a_{ij} = \sum_{(i,j)\in S'} a_{ij} = \sum_{j=1}^{n}\sum_{i=1}^{m} a_{ij}. \qquad \square$$

Here is a summation we will need when we study Taylor Series:

Example.

$$\sum_{i=0}^{N}\sum_{j=0}^{i} a_{ij} = \sum_{j=0}^{N}\sum_{i=j}^{N} a_{ij}$$

Proof: We can rewrite

$$\sum_{i=0}^{N}\sum_{j=0}^{i} a_{ij} = \sum_{(i,j)\in S} a_{ij}$$

where

$$S = \{(i,j)|\ i \text{ and } j \text{ are integers and for each } 0 \le i \le N, 0 \le j \le i\}$$

and

$$\sum_{j=0}^{N}\sum_{i=j}^{N} a_{ij} = \sum_{(i,j)\in S'} a_{ij}$$

where

$$S' = \{(i,j)|\ i \text{ and } j \text{ are integers and for each } 0 \le j \le N, j \le i \le N\}$$

Now we prove that $S' = S$.

- $\subseteq$

  Let $(a,b) \in S$. By definition

  $$\begin{array}{ccccc} 0 & \le & a & \le & N \\ 0 & \le & b & \le & a. \end{array}$$

  By transitivity, this implies

  $$\begin{array}{ccccc} 0 & \le & b & \le & N \\ b & \le & a & \le & N \end{array}$$

  so $(a,b) \in S'$.

- $\supseteq$

  Let $(a,b) \in S'$. Then by definition

  $$\begin{array}{ccccc} 0 & \le & b & \le & N \\ b & \le & a & \le & N \end{array}$$

  Again, we see that

  $$\begin{array}{ccccc} 0 & \le & a & \le & N \\ 0 & \le & b & \le & a. \end{array}$$

  so $(a,b) \in S$. $\qquad\qquad\qquad\qquad\qquad\qquad\qquad\qquad\qquad\qquad\qquad\qquad\square$

7.3 Proving Two Matrices are Equal

In your math career, you are going to see many matrix expressions. For example,

$$\begin{aligned} (A+B)+C &= A+(B+C) \\ (AB)C &= A(BC) \\ A\vec{x} &= \sum_{j=1}^{n} x_j \vec{\alpha}_j \end{aligned}$$

Each of these statements asserts that one matrix is equal to another matrix. But how do you *formally* prove that two matrices are equal?

Back in high school, you proved equality through direct computation. For example, let A, B, C be 2×2 matrices. To prove associativity

$$(AB)C = A(BC)$$

you directly computed

$$(AB)C = \left(\begin{bmatrix} a_{11} & a_{12} \\ a_{21} & a_{22} \end{bmatrix} \begin{bmatrix} b_{11} & b_{12} \\ b_{21} & b_{22} \end{bmatrix} \right) \begin{bmatrix} c_{11} & c_{12} \\ c_{21} & c_{22} \end{bmatrix}$$

$$= \begin{bmatrix} a_{11}b_{11} + a_{12}b_{21} & a_{11}b_{12} + a_{12}b_{22} \\ a_{21}b_{11} + a_{22}b_{21} & a_{21}b_{12} + a_{22}b_{22} \end{bmatrix} \begin{bmatrix} c_{11} & c_{12} \\ c_{21} & c_{22} \end{bmatrix}$$

$$= \begin{bmatrix} c_{11}(a_{11}b_{11} + a_{12}b_{21}) + c_{21}(a_{11}b_{12} + a_{12}b_{22}) & c_{12}(a_{11}b_{11} + a_{12}b_{21}) + c_{22}(a_{11}b_{12} + a_{12}b_{22}) \\ c_{11}(a_{21}b_{11} + a_{22}b_{21}) + c_{21}(a_{21}b_{12} + a_{22}b_{22}) & c_{12}(a_{21}b_{11} + a_{22}b_{21}) + c_{22}(a_{21}b_{12} + a_{22}b_{22}) \end{bmatrix}$$

Then you mindlessly and painfully computed

$$A(BC) = \begin{bmatrix} a_{11} & a_{12} \\ a_{21} & a_{22} \end{bmatrix} \left(\begin{bmatrix} b_{11} & b_{12} \\ b_{21} & b_{22} \end{bmatrix} \begin{bmatrix} c_{11} & c_{12} \\ c_{21} & c_{22} \end{bmatrix} \right)$$

$$= \begin{bmatrix} a_{11} & a_{12} \\ a_{21} & a_{22} \end{bmatrix} \begin{bmatrix} b_{11}c_{11} + b_{12}c_{21} & b_{11}c_{12} + b_{12}c_{22} \\ b_{21}c_{11} + b_{22}c_{21} & b_{21}c_{12} + b_{22}c_{22} \end{bmatrix}$$

$$= \begin{bmatrix} a_{11}(b_{11}c_{11} + b_{12}c_{21}) + a_{12}(b_{21}c_{11} + b_{22}c_{21}) & a_{11}(b_{11}c_{12} + b_{12}c_{22}) + a_{12}(b_{21}c_{12} + b_{22}c_{22}) \\ a_{21}(b_{11}c_{11} + b_{12}c_{21}) + a_{22}(b_{21}c_{11} + b_{22}c_{21}) & a_{21}(b_{11}c_{12} + b_{12}c_{22}) + a_{22}(b_{21}c_{12} + b_{22}c_{22}) \end{bmatrix}$$

After comparing each component and confirming that they matched, you concluded

$$(AB)C = A(BC).$$

But this is a **bone-headed** way to prove the associative law! Here's why:

- This argument does not apply to matrices of *arbitrary*-size.

- It's highly inefficient and, like the Blobfish, horrifyingly ugly to look at.

Instead *we generalize.* We know two matrices are equal if their components are the same. Thus,

To prove two matrices (of the same size) are equal, we have to show for any i, j, the component of A at position ij is the same as the component of B at ij

In our example, we could have just proved that a **single** (but arbitrary) ij component was equal instead of writing out all four components of both matrices.

Now, let's give a *better* (and actual) proof[1] that matrix multiplication is associative.

Example. *For $n \times n$ matrices A, B, C, we have*

$$(AB)C = A(BC)$$

---

[1]For simplicity, lets assume A, B, C are $n \times n$.

Proof: We need to show, for arbitrary component ij, that

$$[(AB)C]_{ij} = [A(BC)]_{ij} \qquad (\star)$$

Starting from the left-hand side, this is the dot product of the i-th row of AB with the j-th column of C:

$$[(AB)C]_{ij} = \sum_{r=1}^{n} [AB]_{ir} c_{rj}.$$

Notice that the ir-entry of AB is the dot product of the i-th row of A with the r-th column of B. Substituting, we get

$$[(AB)C]_{ij} = \sum_{r=1}^{n} \underbrace{\left(\sum_{q=1}^{n} a_{iq} b_{qr} \right)}_{[AB]_{ir}} c_{rj}$$

Pull c_{rj} into the innermost sum to get

$$[(AB)C]_{ij} = \sum_{r=1}^{n} \left(\sum_{q=1}^{n} a_{iq} b_{qr} c_{rj} \right).$$

Now, look at the right-hand side of $(\star)$:

$$[A(BC)]_{ij} = \sum_{r=1}^{n} a_{ir} [BC]_{rj}$$

Again, substitute the dot product definition for $[BC]_{rj}$:

$$\sum_{r=1}^{n} a_{ir} \underbrace{\left(\sum_{q=1}^{n} b_{rq} c_{qj} \right)}_{[BC]_{rj}}$$

and then pull a_{ir} into the innermost sum.

$$[A(BC)]_{ij} = \sum_{r=1}^{n} \left(\sum_{q=1}^{n} a_{ir} b_{rq} c_{qj} \right)$$

Now we have

$$[(AB)C]_{ij} = \sum_{r=1}^{n} \sum_{q=1}^{n} a_{iq} b_{qr} c_{rj}$$

$$[A(BC)]_{ij} = \sum_{r=1}^{n} \sum_{q=1}^{n} a_{ir} b_{rq} c_{qj}.$$

Of course, we can **switch the roles of the dummy variables** in the first equation:

$$[(AB)C]_{ij} = \sum_{q=1}^{n} \sum_{r=1}^{n} a_{ir} b_{rq} c_{qj}$$

$$[A(BC)]_{ij} = \sum_{r=1}^{n} \sum_{q=1}^{n} a_{ir} b_{rq} c_{qj}.$$

Looks better! Now we just have to fix the indexing set. But that's easy: we already proved that we can switch the order of the double summation. Therefore, we can switch the order in the top line:

$$[(AB)C]_{ij} = \sum_{r=1}^{n} \sum_{q=1}^{n} a_{ir} b_{rq} c_{qj}$$

$$[A(BC)]_{ij} = \sum_{r=1}^{n} \sum_{q=1}^{n} a_{ir} b_{rq} c_{qj}$$

Thus,

$$[(AB)C]_{ij} = [A(BC)]_{ij}$$

Since ij was arbitrary, we conclude that all the components agree, hence:

$$(AB)C = A(BC) \qquad \square$$

Notice that we actually had to do some work to prove that matrix multiplication is associative. Associativity is not always obvious! If you ever study Cryptography and elliptic curves, you'll know what I mean. Generally,

> Math Mantra: Just because something is easy to write, DOESN'T MEAN it's easy to prove!

Matrix multiplication is associative. Great. But how about an example of a matrix equality that's more useful for the working mathematician? Sure!

Consider the product of matrix A with vector $\vec{x}$:

$$A\vec{x}.$$

By our definition, you can compute the new vector one element at a time, by dotting the i-th row of A with $\vec{x}$. That's somewhat painful.

The better and more useful idea is to view $A\vec{x}$ as a **linear combination of columns of A**. For example,

$$\begin{bmatrix} 1 & 4 & 7 \\ 2 & 5 & 8 \\ 3 & 6 & 9 \end{bmatrix} \begin{bmatrix} x_1 \\ x_2 \\ x_3 \end{bmatrix} = x_1 \begin{bmatrix} 1 \\ 2 \\ 3 \end{bmatrix} + x_2 \begin{bmatrix} 4 \\ 5 \\ 6 \end{bmatrix} + x_3 \begin{bmatrix} 7 \\ 8 \\ 9 \end{bmatrix}$$

Example. *Let A be an $m \times n$ matrix with columns $\vec{\alpha}_1, \vec{\alpha}_2, \ldots \vec{\alpha}_n$:*

$$\begin{bmatrix} | & | & & | \\ \vec{\alpha}_1 & \vec{\alpha}_2 & \ldots & \vec{\alpha}_n \\ | & | & & | \end{bmatrix}$$

For any vector $\vec{x} \in \mathbb{R}^n$

$$A\vec{x} = x_1\vec{\alpha}_1 + x_2\vec{\alpha}_2 + \ldots + x_n\vec{\alpha}_n = \sum_{j=1}^{n} x_j\vec{\alpha}_j$$

Proof: Consider

$$\sum_{j=1}^{n} x_j\vec{\alpha}_j$$

and look at the i-th component of this sum. This is just sum of all the scaled i-th components of each of the columns:

$$\sum_{j=1}^{n} x_j\vec{\alpha}_j = \begin{bmatrix} x_1 a_{11} \\ \vdots \\ x_1 a_{i1} \\ \vdots \\ x_1 a_{m1} \end{bmatrix} + \begin{bmatrix} x_2 a_{12} \\ \vdots \\ x_2 a_{i2} \\ \vdots \\ x_2 a_{m2} \end{bmatrix} + \ldots + \begin{bmatrix} x_n a_{1n} \\ \vdots \\ x_n a_{in} \\ \vdots \\ x_n a_{mn} \end{bmatrix}$$

Thus,

$$\left[\sum_{j=1}^{n} x_j\vec{\alpha}_j\right]_i = \sum_{j=1}^{n} x_j a_{ij}$$

But the i-th entry of the vector $A\vec{x}$ is, by definition,

$$[A\vec{x}]_i = \sum_{j=1}^{n} a_{ij} x_j$$

Thus,

$$[A\vec{x}]_i = \left[\sum_{j=1}^{n} x_j\vec{\alpha}_j\right]_i$$

Since this is true for every position i,

$$A\vec{x} = \sum_{j=1}^{n} x_j\vec{\alpha}_j \qquad \square$$

Even though this theorem refers to the product of a matrix and vector, we can extend this idea to the product of two matrices. Specifically, we can show that "A distributes across the columns:"

Example. *Let A be an $m \times n$ and B be an $n \times p$ matrix with columns $\vec{\beta}_1, \vec{\beta}_2, \ldots, \vec{\beta}_p$:*

$$B = \begin{bmatrix} | & | & & | \\ \vec{\beta}_1 & \vec{\beta}_2 & \ldots & \vec{\beta}_p \\ | & | & & | \end{bmatrix}$$

Then the columns of AB are then $A\vec{\beta}_1, A\vec{\beta}_2, \ldots, A\vec{\beta}_p$:

$$AB = \begin{bmatrix} | & | & & | \\ A\vec{\beta}_1 & A\vec{\beta}_2 & \ldots & A\vec{\beta}_p \\ | & | & & | \end{bmatrix}$$

Proof: Let's look at

$$\begin{bmatrix} | & | & & | \\ A\vec{\beta}_1 & A\vec{\beta}_2 & \ldots & A\vec{\beta}_p \\ | & | & & | \end{bmatrix}$$

The ij entry of this matrix is the i-th component of $A\vec{\beta}_j$. Writing $A\vec{\beta}_j$ as a linear combination of columns of A,

$$A\vec{\beta}_j = \sum_{r=1}^{N} b_{rj}\vec{\alpha}_r$$

where $\vec{\alpha}_r$ is the r-th column of A. The i-th component is then

$$[A\vec{\beta}_j]_i = \sum_{r=1}^{N} b_{rj}a_{ir}.$$

We also know the ij entry of AB is

$$[AB]_{ij} = \sum_{r=1}^{N} a_{ir}b_{rj}.$$

Thus,

$$[AB]_{ij} = [A\vec{\beta}_j]_i.$$

Since ij was an arbitrary entry, we can conclude

$$AB = \begin{bmatrix} | & | & & | \\ A\vec{\beta}_1 & A\vec{\beta}_2 & \ldots & A\vec{\beta}_p \\ | & | & & | \end{bmatrix} \qquad \square$$

We can also look at the *row analogues* of the last two theorems. These will be vital when we prove the rank theorems.

First up: when we multiply a row vector by a matrix, the result is a linear combination of the rows. For example,

$$\begin{bmatrix} x_1 & x_2 & x_3 \end{bmatrix} \begin{bmatrix} 1 & 2 & 3 \\ 4 & 5 & 6 \\ 7 & 8 & 9 \end{bmatrix} = x_1 \begin{bmatrix} 1 & 2 & 3 \end{bmatrix} + x_2 \begin{bmatrix} 4 & 5 & 6 \end{bmatrix} + x_3 \begin{bmatrix} 7 & 8 & 9 \end{bmatrix}$$

Example. *Let $\vec{x}$ be a $1 \times n$ **row** vector*

$$\vec{x} = \begin{bmatrix} x_1 & x_2 & \ldots & x_n \end{bmatrix}$$

and let B be a $n \times p$ matrix with rows $\vec{B}_1, \vec{B}_2, \ldots, \vec{B}_n$:

$$B = \begin{bmatrix} \underline{\quad\quad} & \vec{B}_1 & \underline{\quad\quad} \\ \underline{\quad\quad} & \vec{B}_2 & \underline{\quad\quad} \\ & \vdots & \\ \underline{\quad\quad} & \vec{B}_n & \underline{\quad\quad} \end{bmatrix}.$$

Then,

$$\vec{x}B = \sum_{i=1}^{n} x_i \vec{B}_i.$$

Proof: Let's look at the j-th entry of

$$\sum_{i=1}^{n} x_i \vec{B}_i$$

Visually, we see that we are isolating the j-th components of the rows:

$$\sum_{i=1}^{n} x_i \vec{B}_i = \begin{cases} \begin{bmatrix} x_1 B_{11} & \ldots & x_1 B_{1j} & \ldots & x_1 B_{1p} \end{bmatrix} \\ \qquad\qquad\qquad + \\ \begin{bmatrix} x_2 B_{21} & \ldots & x_2 B_{2j} & \ldots & x_2 B_{2p} \end{bmatrix} \\ \qquad\qquad\qquad + \\ \qquad\qquad\qquad \vdots \\ \qquad\qquad\qquad + \\ \begin{bmatrix} x_n B_{n1} & \ldots & x_n B_{nj} & \ldots & x_n B_{np} \end{bmatrix} \end{cases}$$

Thus,

$$\left[\sum_{i=1}^{n} x_i \vec{B}_i \right]_j = \sum_{i=1}^{n} x_i B_{ij}$$

But by definition of matrix multiplication,

$$[\vec{x}B]_j = \sum_{i=1}^{n} x_i B_{ij}.$$

This gives us

$$[\vec{x}B]_j = \left[\sum_{i=1}^{n} x_i \vec{B}_i \right]_j$$

Since j was arbitrary, we can conclude

$$\vec{x}B = \sum_{i=1}^{n} x_i \vec{B}_i. \qquad \qquad \square$$

Lastly, we have the "row distributive property:"

Example. *Let A be an $m \times n$ matrix with rows $\vec{A}_1, \vec{A}_2, \ldots, \vec{A}_m$:*

$$A = \begin{bmatrix} \text{---} & \vec{A}_1 & \text{---} \\ \text{---} & \vec{A}_2 & \text{---} \\ & \vdots & \\ \text{---} & \vec{A}_m & \text{---} \end{bmatrix}$$

and let B be an $n \times p$ matrix. Then,

$$AB = \begin{bmatrix} \text{---} & \vec{A}_1 B & \text{---} \\ \text{---} & \vec{A}_2 B & \text{---} \\ & \vdots & \\ \text{---} & \vec{A}_m B & \text{---} \end{bmatrix}$$

Proof: Let's look at

$$\begin{bmatrix} \text{---} & \vec{A}_1 B & \text{---} \\ \text{---} & \vec{A}_2 B & \text{---} \\ & \vdots & \\ \text{---} & \vec{A}_m B & \text{---} \end{bmatrix}$$

The ij entry is the j-th component of the i-th row, $\vec{A}_i B$. Writing $\vec{A}_i B$ as a linear combination of the rows of B,

$$\vec{A}_i B = \sum_{r=1}^{n} a_{ir} \vec{B}_r$$

where $\vec{B}_r$ is the r-th row of B. The j-th component is then

$$[\vec{A}_i B]_j = \sum_{r=1}^{n} a_{ir} b_{rj}.$$

By definition of matrix multiplication, the ij-th entry of AB is

$$[AB]_{ij} = \sum_{r=1}^{n} a_{ir} b_{rj}$$

Thus, we can conclude,

$$AB = \begin{bmatrix} \underline{\quad\quad} & \vec{A_1}B & \underline{\quad\quad} \\ \underline{\quad\quad} & \vec{A_2}B & \underline{\quad\quad} \\ & \vdots & \\ \underline{\quad\quad} & \vec{A_m}B & \underline{\quad\quad} \end{bmatrix}$$

$\square$

7.4 Distances on Matrices

Just as we defined distance between vectors, we can define a *distance function* on matrices. Here, we define a matrix norm:

Definition. *The **norm** of an $m \times n$ matrix A, denoted $\|A\|$, is defined as*

$$\|A\| = \left\| \begin{matrix} a_{11} & a_{12} & \cdots & a_{1n} \\ a_{21} & a_{22} & \cdots & a_{2n} \\ a_{31} & a_{32} & \cdots & a_{3n} \\ \vdots & \vdots & \vdots & \vdots \\ a_{m1} & a_{m2} & \cdots & a_{mn} \end{matrix} \right\| = \sqrt{\sum_{i=1}^{m} \sum_{j=1}^{n} a_{ij}^2}$$

Matrices **inherit** a lot of the distance properties from vectors. Why? The matrix norm is the same as the vector norm: just *unravel* the matrix into a vector.

$$\begin{bmatrix} a_{11} & a_{12} & \cdots & a_{1n} \\ a_{21} & a_{22} & \cdots & a_{2n} \\ a_{31} & a_{32} & \cdots & a_{3n} \\ \vdots & \vdots & \vdots & \vdots \\ a_{m1} & a_{m2} & \cdots & a_{mn} \end{bmatrix} \Rightarrow \begin{bmatrix} a_{11} \\ a_{12} \\ \vdots \\ a_{1n} \\ a_{21} \\ a_{22} \\ \vdots \\ a_{2n} \\ \vdots \\ a_{m1} \\ a_{m2} \\ \vdots \\ a_{mn} \end{bmatrix}$$

In particular, we can prove an important Cauchy-Schwarz-like upper bound:

$$\|A\vec{x}\| \le \|A\| \, \|\vec{x}\|$$

Intuitively, this gives us an upper bound on how much A scales $\vec{x}$ through matrix multiplication. When we hit Multivariable Calculus, this inequality will be our bread and butter.

By the way, be careful! In the above expression, we are using the same symbol $\|\cdot\|$ to denote the *matrix norm* and the *vector norm*:

$$\|A\vec{x}\| \qquad\qquad \|A\| \qquad\qquad \|\vec{x}\|$$
$$\Uparrow \qquad\qquad\quad \Uparrow \qquad\qquad\quad \Uparrow$$
$$\text{Vector Norm} \quad \text{Matrix Norm} \quad \text{Vector Norm}$$

As always,

> ```
> Math Mantra:  Watch out for overloaded notation!
> ```

Theorem. *For any $m \times n$ matrix A and any vector $\vec{x} \in \mathbb{R}^n$,*

$$\|A\vec{x}\| \le \|A\|\,\|\vec{x}\|.$$

Proof: Recall that we proved

$$A\vec{x} = x_1\vec{\alpha}_1 + x_2\vec{\alpha}_2 + \ldots + x_n\vec{\alpha}_n$$

where $\vec{\alpha}_i$ is the i-th column of A. Applying the triangle inequality,

$$\underbrace{\|x_1\vec{\alpha}_1 + x_2\vec{\alpha}_2 + \ldots + x_n\vec{\alpha}_n\|}_{\|A\vec{x}\|} \le \|x_1\vec{\alpha}_1\| + \|x_2\vec{\alpha}_2\| + \ldots + \|x_n\vec{\alpha}_n\|$$

Pulling out the scalars from the norms, we rewrite the upper bound as

$$|x_1|\,\|\vec{\alpha}_1\| + |x_2|\,\|\vec{\alpha}_2\| + \ldots + |x_n|\,\|\vec{\alpha}_n\|$$

Stare at this for a moment: be like Jack in Christmas Town and ask *what's this?*

This is a dot product! Precisely, it is

$$\begin{bmatrix} |x_1| \\ |x_2| \\ \vdots \\ |x_n| \end{bmatrix} \cdot \begin{bmatrix} \|\vec{\alpha}_1\| \\ \|\vec{\alpha}_2\| \\ \vdots \\ \|\vec{\alpha}_n\| \end{bmatrix}$$

And what can we do to dot products? We apply Cauchy-Schwarz! This gives us

$$|x_1|\,\|\vec{\alpha}_1\| + |x_2|\,\|\vec{\alpha}_2\| + \ldots + |x_n|\,\|\vec{\alpha}_n\| \le \underbrace{\sqrt{\sum_{i=1}^{n} x_i^2}}_{\|A\|} \underbrace{\sqrt{\sum_{i=1}^{n} \|\alpha_i\|^2}}_{\|\vec{x}\|}$$

Thus,

$$\|A\vec{x}\| \le \|A\|\,\|\vec{x}\|. \qquad\qquad \square$$

7.5 Importance Behind Matrices: Linear Maps

We developed all this theory for matrices, but we never answered **why** they are important. So here is the big picture:

Any linear map (from $\mathbb{R}^n$ to $\mathbb{R}^m$) is a matrix multiplication!

Theorem. *Let T be a linear map from $\mathbb{R}^n$ to $\mathbb{R}^m$. Then, the function[1] $T(\vec{x})$ can be written as the matrix multiplication:*

$$T(\vec{x}) = A\vec{x}$$

where

$$A = \begin{bmatrix} | & | & & | \\ T(\vec{e_1}) & T(\vec{e_1}) & \dots & T(\vec{e_n}) \\ | & | & & | \end{bmatrix}$$

Proof: Given a vector $\vec{x}$, we can rewrite it in terms of the standard basis vectors $\vec{e_1}, \vec{e_2}, \dots, \vec{e_n}$:

$$\underbrace{\begin{bmatrix} x_1 \\ x_2 \\ \vdots \\ x_n \end{bmatrix}}_{\vec{x}} = x_1 \begin{bmatrix} 1 \\ 0 \\ \vdots \\ 0 \end{bmatrix} + x_2 \begin{bmatrix} 0 \\ 1 \\ \vdots \\ 0 \end{bmatrix} + \dots + x_n \begin{bmatrix} 0 \\ 0 \\ \vdots \\ 1 \end{bmatrix}$$

Condensely,

$$\vec{x} = x_1\vec{e_1} + x_2\vec{e_2} + \dots + x_1\vec{e_n}$$

Applying T,

$$T(\vec{x}) = T(x_1\vec{e_1} + x_2\vec{e_2} + \dots + x_1\vec{e_n})$$

and by linearity,

$$T(\vec{x}) = x_1T(\vec{e_1}) + x_2T(\vec{e_2}) + \dots + x_1T(\vec{e_n})$$

But by our "column distributive property," this can be expressed as the product of a matrix and vector:

$$T(\vec{x}) = A\vec{x}$$

where

$$A = \begin{bmatrix} | & | & & | \\ T(\vec{e_1}) & T(\vec{e_1}) & \dots & T(\vec{e_n}) \\ | & | & & | \end{bmatrix} \qquad \square$$

From this proof, we can make two major observations. The first:

---

[1] Again, be careful about overloaded notation! $T(\vec{x})$ is a function mapping whereas $A\vec{x}$ is a matrix product.

*We have **completely** classified linear maps from $\mathbb{R}^n$ to $\mathbb{R}^m$.*

Notice that all the steps in our proof are completely invertible: we can represent any linear map as a matrix multiplication **and** any matrix multiplication represents a linear map. This means that if we want to talk about linear functions, it's enough just to talk about matrix multiplication!

The second major observation is:

We can determine a linear map T completely by computing its values on the standard basis vectors, namely,

$$T(\vec{e}_1), T(\vec{e}_2), \ldots, T(\vec{e}_n)$$

This is amazing! To represent T, all we need to do is see how T acts on

$$\vec{e}_1, \vec{e}_2, \ldots, \vec{e}_n$$

and then plug in the outputs as the columns of a matrix.

Example. *Let $R_\theta : \mathbb{R}^2 \to \mathbb{R}^2$ denote the mapping that rotates a point in the plane counter-clockwise by θ degrees. Express R_θ as a matrix multiplication.*

Consider the point

$$\begin{bmatrix} 1 \\ 0 \end{bmatrix}$$

Schematically, we can see that R_θ transforms this point:

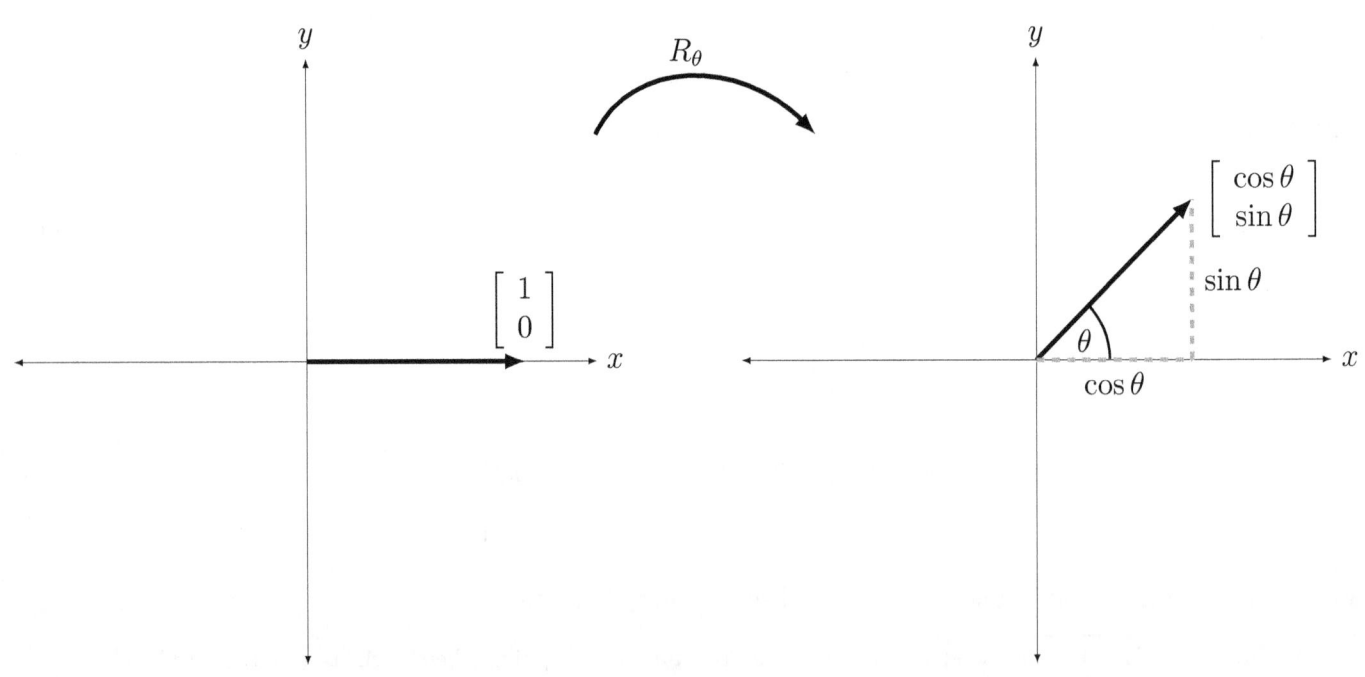

Thus,

$$R_\theta \begin{pmatrix} 1 \\ 0 \end{pmatrix} = \begin{bmatrix} \cos\theta \\ \sin\theta \end{bmatrix}$$

Likewise, we can see how R_θ transforms $\begin{bmatrix} 0 \\ 1 \end{bmatrix}$:

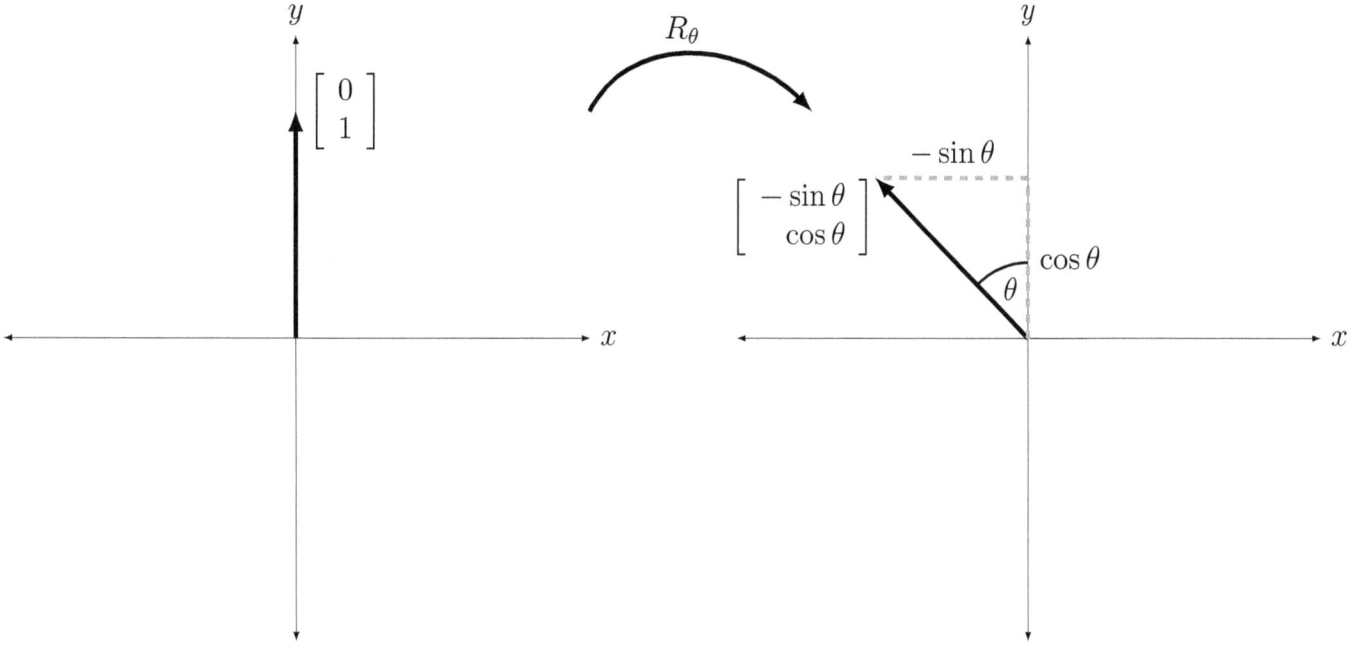

Thus,

$$R_\theta \begin{pmatrix} 0 \\ 1 \end{pmatrix} = \begin{bmatrix} -\sin\theta \\ \cos\theta \end{bmatrix}$$

Applying our preceding theorem,

$$R_\theta \begin{pmatrix} x \\ y \end{pmatrix} = \begin{bmatrix} R_\theta \begin{pmatrix} 1 \\ 0 \end{pmatrix} & R_\theta \begin{pmatrix} 0 \\ 1 \end{pmatrix} \end{bmatrix} \begin{bmatrix} x \\ y \end{bmatrix}$$

$$= \begin{bmatrix} \cos\theta & -\sin\theta \\ \sin\theta & \cos\theta \end{bmatrix} \begin{bmatrix} x \\ y \end{bmatrix}$$

Lecture 8

Row Space, Column Space, Null Space, Oh My!

I'm living in the kernel of a rank-one map.
From my domain, its image looks so blue.
'Cause all I see are zeroes, it's a cruel trap.
But we're a finite simple group of order two.

-Klein Four

Goals: After introducing the row space, column space, and null space, we prove a fundamental relationship about their sizes. Namely, the Rank-Nullity Theorem asserts that the dimensions of the null space and column space sum to the dimension of the domain. The proof of this theorem hinges on our basis theorems from Lecture 6. Lastly, we prove some fundamental rank properties.

8.1   Column Space and Null Space

Most of the material so far has been motivated by the study of linear functions. Particularly, given a linear function f we've seen that the following are all subspaces:

- The *domain* of f.

- The *image* of f.

- The *solution space* of $f(\vec{x}) = \vec{0}$

We also showed, last lecture, that linear functions on $\mathbb{R}^n$ are **directly linked** to matrices: any linear function can be represented as a matrix multiplication

$$A\vec{x}$$

where A is the matrix whose columns are the mapped standard basis vectors. Conversely, any such matrix multiplication represents a linear function.

One question we can ask is,

157

What do the aforementioned subspaces mean when we translate them into the world of matrix multiplication?

- The *domain* of f is all possible $\vec{x}$ that can be plugged into the matrix multiplication:

$$\begin{bmatrix} a_{11} & a_{12} & \cdots & a_{1n} \\ a_{21} & a_{22} & \cdots & a_{2n} \\ \vdots & \vdots & \ddots & \vdots \\ a_{m1} & a_{m2} & \cdots & a_{mn} \end{bmatrix} \begin{bmatrix} x_1 \\ x_1 \\ \vdots \\ x_n \end{bmatrix}$$

Namely, it is the space $\mathbb{R}^n$.

- The *image* of f is the *output* from plugging in all possible $\vec{x}$. But we showed that a matrix multiplied by a vector simply outputs a linear combination of the columns:

$$\begin{bmatrix} | & | & & | \\ \vec{\alpha}_1 & \vec{\alpha}_2 & \cdots & \vec{\alpha}_n \\ | & | & & | \end{bmatrix} \begin{bmatrix} x_1 \\ x_2 \\ \vdots \\ x_n \end{bmatrix} = x_1 \begin{bmatrix} | \\ \vec{\alpha}_1 \\ | \end{bmatrix} + x_2 \begin{bmatrix} | \\ \vec{\alpha}_1 \\ | \end{bmatrix} + \ldots + x_n \begin{bmatrix} | \\ \vec{\alpha}_1 \\ | \end{bmatrix}$$

Thus, the image is really the *span* of the columns of A. We call this span the *column space*.

- The *solution space* of $f(\vec{x}) = \vec{0}$ is the set of vectors $\vec{x}$ that A multiplies to $\vec{0}$:

$$\begin{bmatrix} a_{11} & a_{12} & \cdots & a_{1n} \\ a_{21} & a_{22} & \cdots & a_{2n} \\ \vdots & \vdots & \ddots & \vdots \\ a_{m1} & a_{m2} & \cdots & a_{mn} \end{bmatrix} \begin{bmatrix} x_1 \\ x_2 \\ \vdots \\ x_n \end{bmatrix} = \begin{bmatrix} 0 \\ 0 \\ \vdots \\ 0 \end{bmatrix}$$

We call this the *null space* of A.

Formally, we define:

Definition. *Let A be an $m \times n$ matrix. Then,*

- *The **domain** of A is $\mathbb{R}^n$.*

- *The **column space** of A is the subspace of $\mathbb{R}^m$:*

$$C(A) = \text{span}\{\vec{\alpha}_1, \vec{\alpha}_2, \ldots, \vec{\alpha}_n\}$$

where the $\vec{\alpha}_i$ are the columns of A.

- The **null space** of A is the subspace

$$N(A) = \left\{ \vec{x} \in \mathbb{R}^n \,\middle|\, A\vec{x} = \vec{0} \right\}.$$

We call the dimension of $C(A)$ the **column rank**, and the dimension of $N(A)$ the **nullity**.

One question we can ask is

Is there any relationship between the domain, the column space, and the null space?

For starters, since the null space is contained in the domain $\mathbb{R}^n$ and the image is contained in the space $\mathbb{R}^m$,

$$\begin{aligned} \dim N(A) &\leq n \\ \dim C(A) &\leq m \end{aligned}$$

But these are pretty obvious (and pretty lame) inequalities. How about something more exciting?

Fortunately, we have an **incredible** result, the *Rank-Nullity Theorem*, which precisely relates the dimensions of these three subspaces. In fact, Gilbert Strang refers to this theorem as the first part of the **Fundamental Theorem of Linear Algebra.**[1]

8.2   Rank-Nullity Theorem

This is the most important application of the Basis Theorem. Namely, it gives you a direct relationship between the "size" of the the domain, the column space, and the null space:

The "size" of the domain is the sum of the "sizes" of the column space and the null space.[2]

Precisely, for any $m \times n$ matrix A,

$$\dim C(A) + \dim N(A) = n$$

One interpretation is that, after mapping by A, every vector in the domain $\mathbb{R}^n$ is either "killed" and sent to zero or used to form the column space:

---

[1]I highly encourage you to google his MAA article with the same name. It requires only a modest Linear Algebra background and has excellent illustrations!

[2]Since kernel is a synonym for null space, the lyrics at the beginning of this lecture imply that he's living in a very big null space!

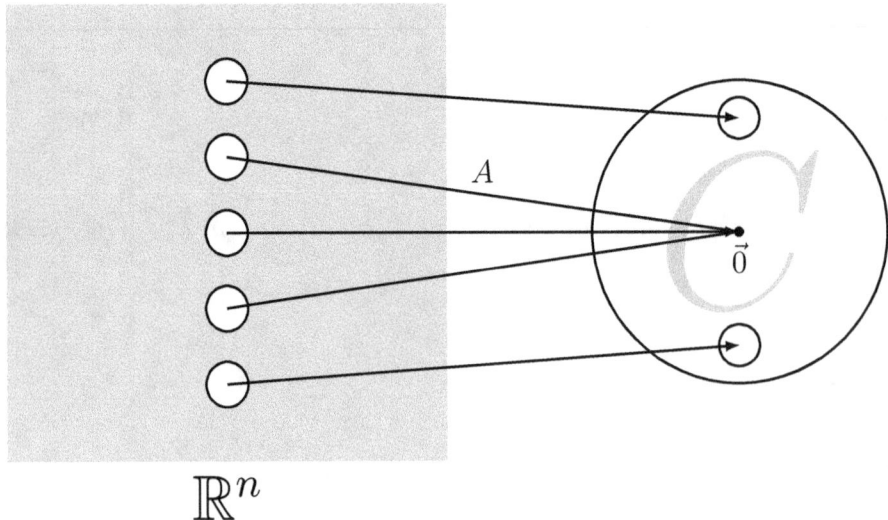

$$\mathbb{R}^n$$

If we add the "size" of the collection of vectors "killed" and the "size" of the collection of vectors used to build the column space, we would get the "size" of the full domain.

Or, as a bar-nalogy, imagine pouring tequila into a shot-glass. Chances are, you are going to spill some of that onto the bar mat. If you recombine the shot and the spilled tequila (via a sponge), the conglomerated booze is the same as the original amount poured.

But why is this theorem worth being called the *first part of the Fundamental Theorem of Linear Algebra?*

1. **The Rank-Nullity Theorem is a counting result.** Notice that n is a given fixed constant. This means that we can directly calculate the dimension of the null space given the dimension of the column space and vice versa. That's pretty neat! Moreover, we can use the Rank-Nullity Theorem to prove certain linear mappings are impossible: otherwise things "wouldn't add up." For example, we can never have a linear map with domain $\mathbb{R}^2$ and column space $C(A) = \mathbb{R}^3$: otherwise, by the Rank-Nullity Theorem:

$$3 + \dim N(A) = 2$$

   so $\dim N(A) = $ -1, which is absurd.

2. **The Rank-Nullity Theorem is an existence result.** If we know the dimension of subspace is a certain number k, this gives us a basis of k vectors to work with! For example, if we know $n = 5$ and $\dim C(A) = 2$, then

$$2 + \dim N(A) = 5$$

   Thus, $\dim N(A) = 3$. Even though we have no clue how to calculate the null space, we still know it has some basis

$$\vec{v}_1, \vec{v}_2, \vec{v}_3$$

   that we can work with.[1]

---
[1]You are going to see this trick *a lot* in the next lecture, specifically when deriving limit values.

3. **The Rank-Nullity Theorem is going to pop up *many* times.** We are going to use it to prove relationships between rows and columns, as well as facts about invertibility of matrices. The Rank-Nullity Theorem even shows up in **Math 52H** and **Math 53H**, for example, in the study of Jordan Canonical forms.

But how do you prove the Rank-Nullity Theorem?

First notice that $N(A)$ is contained in the domain, $\mathbb{R}^n$:

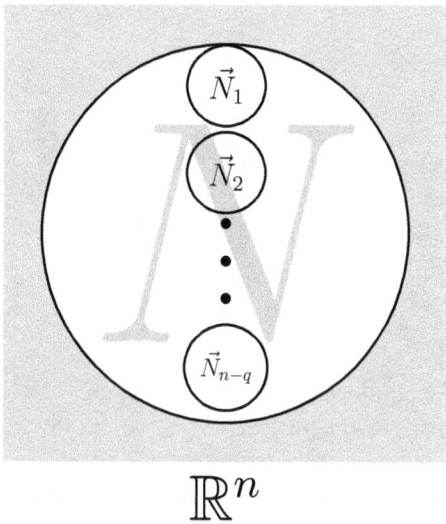

$$\mathbb{R}^n$$

By basis extension theorem, we can extend the null space basis

$$\vec{N}_1, \vec{N}_2, \ldots, \vec{N}_{n-q}$$

to a full n-vector basis for $\mathbb{R}^n$:

$$\vec{N}_1, \vec{N}_2, \ldots \vec{N}_{n-q}, \vec{x}_1, \vec{x}_2, \ldots \vec{x}_q.$$

Remarkably, we can prove that the *image of the extension vectors* under A

$$A\vec{x}_1, A\vec{x}_2, \ldots, A\vec{x}_q$$

is a basis for the column space:

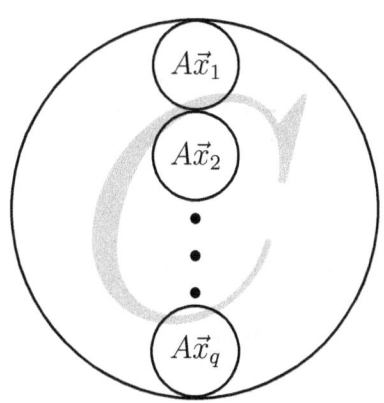

Thus, the sum of the dimensions of the null space and the column space equals n, which is the dimension of the domain $\mathbb{R}^n$:

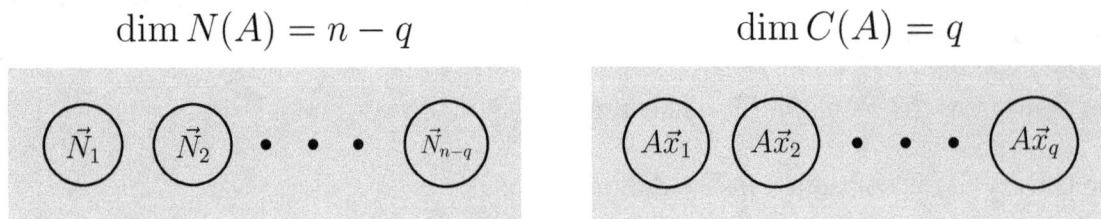

$$\dim N(A) = n - q \qquad\qquad \dim C(A) = q$$

Easy as π.

Theorem (Rank-Nullity Theorem). *For any $m \times n$ matrix A,*

$$\dim C(A) + \dim N(A) = n$$

Proof Summary:

- Extend null space basis to a full basis for $\mathbb{R}^n$.

- Show the image of the extension vectors under A is a basis for the column space.

  - *Linear Independence:*

    * Suppose not. Then there exists a non-trivial combination of $A(x_i)$ whose sum is $\vec{0}$
    * Apply linearity of A to show A maps a non-trivial combination of $\vec{x}_i$ to $\vec{0}$.
    * Therefore, the combination is in the null space.
    * This contradicts linear independence of our original basis for $\mathbb{R}^n$.

  - *Spanning:*

    * span $\{A\vec{x}_1, A\vec{x}_2, \ldots, A\vec{x}_n\} \subseteq C(A)$
      Definition.

    * $C(A) \supseteq \text{span}\{A\vec{x}_1, A\vec{x}_2, \ldots, A\vec{x}_n\}$
      For $\vec{y} \in C(A)$, $A\vec{x} = \vec{y}$. Expand $\vec{x}$ in terms of original basis.

- Conclude
$$\underbrace{\dim C(A)}_{q} + \underbrace{\dim N(A)}_{n-q} = n.$$

Proof: By the basis theorem, we know that $N(A)$ has a basis. Moreover, since $N(A) \subseteq \mathbb{R}^n$,

$$\dim N(A) \leq n.$$

Therefore, we can assume $\dim N(A) = n - q$ vectors where q is some non-negative integer:

$$\vec{N}_1, \vec{N}_2, \ldots \vec{N}_{n-q}.$$

Applying extension theorem, extend $N(A)$ to a full n-dimensional basis[1] for $\mathbb{R}^n$:

$$\vec{N}_1, \vec{N}_2, \ldots \vec{N}_{n-q}, \vec{x}_1, \vec{x}_2, \ldots \vec{x}_q.$$

I claim that the image of the extension vectors under A

$$A\vec{x}_1, A\vec{x}_2, \ldots, A\vec{x}_q$$

gives us a basis for $C(A)$. This would complete the proof since

$$\dim C(A) = q$$

and thus

$$\underbrace{\dim C(A)}_{q} + \underbrace{\dim N(A)}_{n-q} = n.$$

So let's check the definition of a basis!

- *Linearly independent.*

  Suppose

  $$A\vec{x}_1, A\vec{x}_2, \ldots, A\vec{x}_q$$

  is not linearly independent. Then we have some non-trivial combination

  $$\alpha_1 A(\vec{x}_1) + \alpha_2 A(\vec{x}_2) + \ldots + \alpha_q A(\vec{x}_q) = \vec{0}$$

  By linearity of matrix multiplication,

  $$A\left(\alpha_1 \vec{x}_1 + \alpha_2 \vec{x}_2 + \ldots + \alpha_q \vec{x}_q\right) = 0.$$

  Thus,

  $$\alpha_1 \vec{x}_1 + \alpha_2 \vec{x}_2 + \ldots + \alpha_q \vec{x}_q$$

  is in the null space of A. Writing this vector in terms of the null space basis,

  $$\alpha_1 \vec{x}_1 + \alpha_2 \vec{x}_2 + \ldots + \alpha_q \vec{x}_q = \beta_1 \vec{N}_1 + \beta_2 \vec{N}_2 + \ldots + \beta_{n-q} \vec{N}_{n-q}.$$

  This gives us a non-trivial combination for $\vec{0}$:

  $$\alpha_1 \vec{x}_1 + \alpha_2 \vec{x}_2 + \ldots + \alpha_k \vec{x}_q - \beta_1 \vec{N}_1 - \beta_2 \vec{N}_2 - \ldots - \beta_{n-q} \vec{N}_{n-q} = \vec{0}.$$

  However, the above vectors are elements of our original basis

  $$\vec{N}_1, \vec{N}_2, \ldots \vec{N}_{n-q}, \vec{x}_1, \vec{x}_2, \ldots \vec{x}_q$$

  for $\mathbb{R}^n$, directly contradicting linear independence! Thus,

  $$A\vec{x}_1, A\vec{x}_2, \ldots, A\vec{x}_q$$

  are linearly independent.

---

[1]Often, students confuse m and n in the statement of the Rank-Nullity Theorem. Think *domain*. If you insist on a mnemonic: extend the <u>n</u>ull space.

- span $\{A\vec{x}_1, A\vec{x}_2, \ldots, A\vec{x}_q\} = C(A)$

  - $\subseteq$

    This is a freebie since the image of specific vectors (as well as their combinations) are automatically in the column space.

  - $\supseteq$

    Let $\vec{c} \in C(A)$. By definition of the column space, we can find a $\vec{y} \in \mathbb{R}^n$ such that

    $$A\vec{y} = \vec{c} \tag{$\star$}$$

Expand $\vec{y}$ in terms of our basis for $\mathbb{R}^n$

$$\vec{y} = \alpha_1 \vec{N}_1 + \alpha_2 \vec{N}_2 + \ldots + \alpha_{n-q} \vec{N}_{n-q} + \beta_1 \vec{x}_1 + \beta_2 \vec{x}_2 + \ldots + \beta_q \vec{x}_q$$

to get

$$A\vec{y} = A\left(\alpha_1 \vec{N}_1 + \alpha_2 \vec{N}_2 + \ldots + \alpha_{n-q} \vec{N}_{n-q} + \beta_1 \vec{x}_1 + \beta_2 \vec{x}_2 + \ldots + \beta_q \vec{x}_q\right).$$

Distribute

$$A\vec{y} = \alpha_1 \underbrace{A\vec{N}_1}_{=0} + \alpha_2 \underbrace{A\vec{N}_2}_{=0} + \ldots + \alpha_{n-q} \underbrace{A\vec{N}_{n-q}}_{=0} + \beta_1 A\vec{x}_1 + \beta_2 A\vec{x}_2 + \ldots + \beta_q A\vec{x}_q$$

and use the fact that each $\vec{N}_i$ is in the null space to get

$$A\vec{y} = \beta_1 A\vec{x}_1 + \beta_2 A\vec{x}_2 + \ldots + \beta_q A\vec{x}_q$$

Plugging this into the (LHS) of $(\star)$,

$$\underbrace{\beta_1 A\vec{x}_1 + \beta_2 A\vec{x}_2 + \ldots + \beta_q A\vec{x}_q}_{A\vec{y}} = \vec{c}$$

By definition, this means

$$\vec{c} \in \text{span}\,\{A\vec{x}_1, A\vec{x}_2, \ldots, A\vec{x}_q\} = C(A)$$

Since we proved the claim, we conclude

$$\underbrace{\dim C(A)}_{q} + \underbrace{\dim N(A)}_{n-q} = n. \qquad \square$$

8.3   Row Space

Notice that the preceding theorem is called the *Rank-Nullity Theorem*, not the *Column Rank-Nullity Theorem*. This is because *rank* is more special than column rank (though they are the same number)! We are going to **upgrade** our Rank-Nullity one step further. Or, to quote Elzar,

We are going to knock it up a notch!

Precisely, we are going to show that the dimension of the *column space* is the same as the dimension of the *row space*. And this number will be called the *rank*.

As a first guess, you may think that:

The row space is the span of the rows of A.

This is **almost** correct. However, we only defined spans for *column* vectors!

To fix this, we make a definition:

Definition. *Given an $m \times n$ matrix*

$$A = \begin{bmatrix} a_{11} & a_{12} & \cdots & a_{1n} \\ a_{21} & a_{22} & \cdots & a_{2n} \\ \vdots & \vdots & \ddots & \vdots \\ a_{m1} & a_{m2} & \cdots & a_{mn} \end{bmatrix}$$

the ***transpose*** *of A is the $n \times m$ matrix whose ij-component is the ji-component of A:*

$$A^T = \begin{bmatrix} a_{11} & a_{21} & \cdots & a_{m1} \\ a_{12} & a_{22} & \cdots & a_{m2} \\ \vdots & \vdots & \ddots & \vdots \\ a_{1n} & a_{2n} & \cdots & a_{mn} \end{bmatrix}$$

Transposes will be extremely important in **Lecture 30**. For now, we will only transpose row vectors. In this case, we are simply propping a row vector as a column vector:

$$\vec{x} = \begin{bmatrix} x_1 & x_2 & \cdots & x_n \end{bmatrix} \qquad \vec{x}^T = \begin{bmatrix} x_1 \\ x_2 \\ \vdots \\ x_n \end{bmatrix}$$

The **correct** definition of the row space is:

The row space is the span of the ***transposed*** *rows of A.*

Precisely,

Definition. *Let A be an $m \times n$ matrix with rows $\vec{A}_1, \vec{A}_2, \ldots, \vec{A}_m$:*

$$A = \begin{bmatrix} \underline{\quad} & \vec{A}_1 & \underline{\quad} \\ \underline{\quad} & \vec{A}_2 & \underline{\quad} \\ & \vdots & \\ \underline{\quad} & \vec{A}_m & \underline{\quad} \end{bmatrix}$$

*The **row space** of A is the subspace of $\mathbb{R}^n$:*

$$R(A) = \operatorname{span}\left\{\vec{A}_1^T, \vec{A}_2^T, \ldots, \vec{A}_m^T\right\}$$

For example, given the matrix

$$A = \begin{bmatrix} 1 & 2 & 3 & 4 & 5 \\ 11 & 12 & 13 & 14 & 15 \end{bmatrix}$$

the row space of A is

$$R(A) = \operatorname{span}\left\{\begin{bmatrix} 1 \\ 2 \\ 3 \\ 4 \\ 5 \end{bmatrix}, \begin{bmatrix} 11 \\ 12 \\ 13 \\ 14 \\ 15 \end{bmatrix}\right\}$$

Using the Rank-Nullity Theorem, we can prove the awesome fact that the dimension of the row space equals the dimension of the column space:

$$\dim R(A) = \dim C(A).$$

This is not obvious! You would think that the rows and columns would have nothing to do with each other. Indeed, the row space and column space are **subsets of different spaces** ($\mathbb{R}^n$ and $\mathbb{R}^m$, respectively). But it turns out that their bases have the same size!

And *once we prove this equality*, we will tear down the walls of column and row discrimination and just say *rank*:

Definition. *Let A be a $m \times n$ matrix. The **rank** of A is defined as*

$$\operatorname{rank} A = \dim C(A) = \dim R(A)$$

Then, the Rank-Nullity Theorem becomes

$$\operatorname{rank} A + \dim N(A) = n.$$

In many linear algebra books, this result is the crowning jewel of the chapter because it has a ton of applications, both theoretical and non-theoretical. Unfortunately, we are going to save the proof that

$$\dim C(A) = \dim R(A)$$

for next week. For now, let's assume the result and get some practice proving rank properties.

8.4 Proving Rank Properties

Here are three basic rank properties that you're expected to prove, assuming A is an $m \times n$ matrix:

$$\text{rank}(A + B) \leq \text{rank } A + \text{rank } B$$
$$\text{rank } A \leq \min\{m, n\}$$
$$\text{rank}(AB) \leq \min\{\text{rank } A, \text{rank } B\}$$

But how do you even begin?

The first rule of Math Fight Club is:

> Math Mantra: Don't let any math statement scare you. Go back to the definition and view the statement in simpler terms!

When you expand the definition of rank, you are going to realize it's just all the basis stuff you've been practicing until this point! The one *minor* difference is that the rank takes a **matrix** an input whereas the dimension takes a subspace as input!

The second rule of Math Fight Club is:

> Math Mantra: Don't even ATTEMPT to prove a math theorem until you are comfortable with all the definitions used!

Go to the Flomo catwalks. Or Roble Field. Or the Synergy garden (if you're into that stuff). Somewhere. Just stare at the sky and think about the definitions. Think about what they *mean*. Then, only when you are ready, attempt the proof. In the words of Professor Simon,

> ***Mull it over*** *until you feel comfortable with the definition. Write out basic cases.*
> *Then write out the extreme cases.*

Example. *Let A and B be $m \times n$ matrices. Then,*

$$\text{rank } (A + B) \leq \text{rank } A + \text{rank } B$$

Proof: Let

$$A = \begin{bmatrix} | & | & & | \\ \vec{\alpha}_1 & \vec{\alpha}_2 & \dots & \vec{\alpha}_n \\ | & | & & | \end{bmatrix} \qquad B = \begin{bmatrix} | & | & & | \\ \vec{\beta}_1 & \vec{\beta}_2 & \dots & \vec{\beta}_n \\ | & | & & | \end{bmatrix}.$$

Then,

$$A + B = \begin{bmatrix} | & | & & | \\ \vec{\alpha}_1 + \vec{\beta}_1 & \vec{\alpha}_2 + \vec{\beta}_2 & \ldots \vec{\alpha}_n + \vec{\beta}_n \\ | & | & & | \end{bmatrix}.$$

Because rank is the dimension of the column space, let's study the span of the columns of $A + B$:

$$C(A + B) = \text{span}\left\{ \vec{\alpha}_1 + \vec{\beta}_1, \vec{\alpha}_2 + \vec{\beta}_2, \ldots, \vec{\alpha}_n + \vec{\beta}_n \right\}.$$

Since any linear combination of these columns

$$c_1(\vec{\alpha}_1 + \vec{\beta}_1) + c_2(\vec{\alpha}_2 + \vec{\beta}_2) + \ldots + c_n(\vec{\alpha}_n + \vec{\beta}_n)$$

is just

$$c_1\vec{\alpha}_1 + c_2\vec{\alpha}_2 + \ldots + c_n\vec{\alpha}_n + c_1\vec{\beta}_1 + c_2\vec{\beta}_2 + \ldots + c_n\vec{\beta}_n$$

we conclude that

$$\text{span}\left\{ \vec{\alpha}_1 + \vec{\beta}_1, \vec{\alpha}_2 + \vec{\beta}_2, \ldots, \vec{\alpha}_n + \vec{\beta}_n \right\} \subseteq \text{span}\left\{ \vec{\alpha}_1, \vec{\alpha}_2, \ldots, \vec{\alpha}_n, \vec{\beta}_1, \vec{\beta}_2, \ldots, \vec{\beta}_n \right\}.$$

By our span theorems, we can collapse a linearly dependent spanning set into a basis. Reordering if necessary,

$$\vec{\alpha}_1, \vec{\alpha}_2, \quad \ldots, \quad \vec{\alpha}_n$$
$$\vec{\beta}_1, \vec{\beta}_2, \quad \ldots, \quad \vec{\beta}_n$$

collapses into bases

$$\vec{\alpha}_1, \vec{\alpha}_2, \quad \ldots, \quad \vec{\alpha}_{\text{rank } A}$$
$$\vec{\beta}_1, \vec{\beta}_2, \quad \ldots, \quad \vec{\beta}_{\text{rank } B}$$

respectively. Therefore,

$$\text{span}\left\{ \vec{\alpha}_1, \vec{\alpha}_2, \ldots, \vec{\alpha}_n, \vec{\beta}_1, \vec{\beta}_2, \ldots, \vec{\beta}_n \right\} = \text{span}\left\{ \vec{\alpha}_1, \vec{\alpha}_2, \ldots, \vec{\alpha}_{\text{rank } A}, \vec{\beta}_1, \vec{\beta}_2, \ldots, \vec{\beta}_{\text{rank } B} \right\}$$

making our set inclusion

$$\text{span}\left\{ \vec{\alpha}_1 + \vec{\beta}_1, \vec{\alpha}_2 + \vec{\beta}_2, \ldots, \vec{\alpha}_n + \vec{\beta}_n \right\} \subseteq \text{span}\left\{ \vec{\alpha}_1, \vec{\alpha}_2, \ldots, \vec{\alpha}_{\text{rank } A}, \vec{\beta}_1, \vec{\beta}_2, \ldots, \vec{\beta}_{\text{rank } B} \right\}.$$

Even though we don't know the dimension of the right hand side, in the **worst-case scenario**, its dimension is rank A + rank B (this happens if all the vectors on the right hand side are linearly independent). Thus,

$$\dim C(A + B) \leq \text{rank } A + \text{rank } B$$

or by definition of rank,

$$\text{rank}(A + B) \leq \text{rank } A + \text{rank } B. \qquad \square$$

The next example is an immediate[1] corollary of our definition of rank.

---

[1]**Do not** submit this proof on Homework 2! That particular exercise defines rank $A = \dim C(A)$. **You cannot assume** $\dim C(A) = \dim R(A)$.

Theorem. *Let A be an $m \times n$ matrix. Then*

$$\text{rank } A \leq \min\{m, n\}$$

Proof: The column space is contained in $\mathbb{R}^m$, so

$$\underbrace{\text{rank } A}_{\dim C(A)} \leq m$$

Likewise, the row space is contained in $\mathbb{R}^n$:

$$\underbrace{\text{rank } A}_{\dim R(A)} \leq n.$$

Thus, rank A is less than or equal to the smaller of m and n:

$$\text{rank } A \leq \min\{m, n\} \qquad \square$$

The last property is about the rank of a product of matrices.

Theorem. *For an $m \times n$ matrix A and $n \times p$ matrix B,*

$$\text{rank } AB \leq \min\left\{\text{rank } A, \text{rank } B\right\}$$

Proof: We need to prove two inequalities separately

- rank $AB \leq$ rank A.

Let's look at the columns of AB. Writing

$$B = \begin{bmatrix} \vert & \vert & & \vert \\ \vec{\beta}_1 & \vec{\beta}_2 & \cdots & \vec{\beta}_p \\ \vert & \vert & & \vert \end{bmatrix},$$

apply the "column distributive law" to get

$$AB = \begin{bmatrix} \vert & \vert & & \vert \\ A\vec{\beta}_1 & A\vec{\beta}_2 & \cdots & A\vec{\beta}_p \\ \vert & \vert & & \vert \end{bmatrix}.$$

By our matrix multiplication properties, each column of AB is a *linear combination* of the columns $\alpha_1, \vec{\alpha}_2, \ldots, \vec{\alpha}_n$ of A:

$$A\vec{\beta}_j = \sum_{r=1}^{n} b_{rj}\vec{\alpha}_r.$$

Therefore, any linear combination of the columns of AB is also a linear combination of the columns of A, hence:

$$\underbrace{\mathrm{span}\left\{A\vec{\beta}_1, A\vec{\beta}_2, \ldots, A\vec{\beta}_p\right\}}_{C(AB)} \subseteq \underbrace{\mathrm{span}\left\{\vec{\alpha}_1, \vec{\alpha}_2, \ldots, \vec{\alpha}_n\right\}}_{C(A)}.$$

Thus,

$$\mathrm{rank}\ AB \leq \mathrm{rank}\ A$$

- rank $AB \leq$ rank B

Look at the rows of A:

$$A = \begin{bmatrix} \rule{1cm}{0.4pt} & \vec{A}_1 & \rule{1cm}{0.4pt} \\ \rule{1cm}{0.4pt} & \vec{A}_2 & \rule{1cm}{0.4pt} \\ & \vdots & \\ \rule{1cm}{0.4pt} & \vec{A}_m & \rule{1cm}{0.4pt} \end{bmatrix}$$

From our matrix multiplication properties, we know that each row of AB is

$$AB = \begin{bmatrix} \rule{1cm}{0.4pt} & \vec{A}_1 B & \rule{1cm}{0.4pt} \\ \rule{1cm}{0.4pt} & \vec{A}_2 B & \rule{1cm}{0.4pt} \\ & \vdots & \\ \rule{1cm}{0.4pt} & \vec{A}_m B & \rule{1cm}{0.4pt} \end{bmatrix}$$

Moreover, each row of AB is a *linear combination* of the rows $\vec{B}_1, \vec{B}_2, \ldots, \vec{B}_n$ of B:

$$\vec{A}_i B = \sum_{r=1}^{n} a_{ir}\vec{B}_r$$

The row space of AB is therefore contained in the row space of B:

$$\underbrace{\mathrm{span}\left\{\left(\vec{A}_1 B\right)^T, \left(\vec{A}_2 B\right)^T, \ldots, \left(\vec{A}_m B\right)^T\right\}}_{R(AB)} \subseteq \underbrace{\mathrm{span}\left\{\vec{B}_1^T, \vec{B}_2^T, \ldots \vec{B}_n^T\right\}}_{R(B)}$$

This implies the dimension of the row space of AB is less than or equal to the dimension of the row space of B:

$$\mathrm{rank}\ AB \leq \mathrm{rank}\ B.$$

Since

$$\begin{aligned} \mathrm{rank}\ AB &\leq \mathrm{rank}\ A \\ \mathrm{rank}\ AB &\leq \mathrm{rank}\ B \end{aligned}$$

rank AB is bounded by the smaller of rank A and rank B:

$$\text{rank } AB \leq \min\{\text{rank } A, \text{rank } B\}. \qquad \square$$

By the way, now that we proved these *rank* theorems, you should be asking yourself why you should you care.

These properties are important *both* theoretically and practically. Theoretically,

$$\text{rank}(A + B) \leq \text{rank } A + \text{rank } B$$

says that adding two matrices cannot create a matrix with a rank greater than the sum of the ranks of the original two matrices. And,

$$\text{rank } A \leq \min\{m, n\}$$

tells us if A is $m \times n$ with $m < n$, the columns must always be linearly dependent. Likewise, if $m > n$, then the rows are always linearly dependent. Finally,

$$\text{rank } AB \leq \min\{\text{rank } A, \text{rank } B\}$$

tells us that when we multiply two matrices, we cannot create a matrix whose rank is greater than the ranks of the two original matrices.

As for practical applications, unfortunately, you won't see any real-world application of rank formulas in Math 51H. However, I highly recommend taking

EE263: Linear Dynamical Systems.

It's an excellent course. In fact, Professor Boyd and his crack team of TA's literally scoured the globe for every important real-world application of linear algebra, assembling a *massive* treasure trove of 100+ problems. Even though it is a graduate level class, it is a lot easier than the H-Series. And every engineer agrees that it is a **must-take** course.

New Notation

Symbol	Reading	Example	Example Translation
$C(A)$	The column space of A	$C(A) = \mathbb{R}^n$	The column space of A is $\mathbb{R}^n$
$N(A)$	The null space of A	$N(A) = \{\vec{0}\}$	The null space of A is the set containing the zero vector
$R(A)$	The row space of A	$\dim R(A) = \dim C(A)$	The dimension of the row space equals the dimension of the column space.
$\vec{x}^T$	$\vec{x}$ transpose	$\begin{bmatrix} 1 & 0 & 0 \end{bmatrix}^T = \vec{e}_1$	The transpose of the vector $\begin{bmatrix} 1 & 0 & 0 \end{bmatrix}$ is the first standard basis vector.
rank A	The rank of A	$\text{rank } A = \dim R(A)$	The rank of the A is equal to the dimension of the row space of A.

Lecture 9

The Sky's the Limit

The ε-N definition took a hundred years to develop.
Yet you are expected to learn it in less than twenty minutes!

- Leon Simon

Goals: We introduce the central notion of analysis, limits. Because this topic is so fundamental, I begin by giving the intuition behind the definition and explaining triple quantifiers. Then, I devote the rest of the lecture to numerous examples, to give you practice with ϵ-N proofs.

9.1   Capturing Closeness

The essence of Calculus is the study of closeness, or rather, the

$$Limit \ of \ processes.$$

For example,

- The derivative is the *limit* of the slopes of secant lines.

- The integral is the *limit* of area approximations.

- An "infinite" sum is the *limit* of partial sums.

But what precisely is a *limit*?

In seven years of teaching, I have never seen a high school teacher actually teach the limit definition correctly. Why? Because kids are squeamish about it. And teachers are squeamish[1] about it. So instead of getting a definition, you are taught to *plug and chug*.

For example, to calculate the limit of the sequence

$$a_n = \frac{n}{e^n}$$

---

[1]In their defence, not even Newton had the correct definition of limit. And that guy wrote calculus in one summer!

you plugged in 10, 50, 100, 200, and saw that the sequence gets closer to 0 as n approaches infinity:

$$a_{10} \approx .0005$$
$$a_{50} \approx 9.16 \times 10^{-21}$$
$$a_{100} \approx 3.72 \times 10^{-42}$$
$$a_{200} \approx 2.76 \times 10^{-85}$$

Then you concluded that a_n converges to 0 as n approaches infinity:

$$a_n \to 0.$$

However,

Math Mantra: You cannot conclude[1] a general result by just plugging in a few numbers!

Consider the sequence
$$a_n = \begin{cases} 0 & : \text{ if } n \text{ is a multiple of 10} \\ n & : \text{ else} \end{cases}$$

Plugging in 10, 50, 100, 200 yields 0:

$$a_{10} = 0$$
$$a_{50} = 0$$
$$a_{100} = 0$$
$$a_{200} = 0$$

But this sequence

$$1, 2, 3, 4, 5, 6, 7, 8, 9, \mathbf{0}, 11 \ldots, 101, 102, 103, 104, 105, 106, 107, 108, 109, \mathbf{0}, 111 \ldots$$

does not converge to 0!

9.2   Intuition for Limits

Before I give you a *rigorous definition* of the limit of a sequence, let's make an analogy. Consider the scenario of **a man shooting arrows**. The n-th shot is the n-th term. Moreover, each shot lands on the number line.

For example, the sequence (a_n) where
$$a_n = \frac{n-1}{n}$$

can be visualized as shots getting closer to 1:

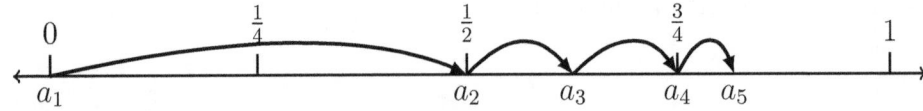

---

[1]Always remember the $x^2 + x + 41$ example. The secret of the universe is not 42; it's 41

Of course, we could be dealing with Sir Robert Locksley of Nottingham[1] (aka Robin Hood): he could hit the same spot twice! Consider the sequence

$$a_n = (-1)^n$$

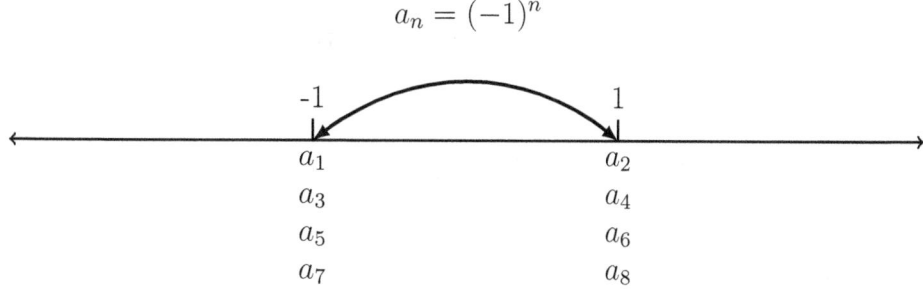

This man's a boss. He goes back and forth hitting the -1, 1 marks every time!

As a *first attempt* at understanding limits, we guess that:

The limit is the bullseye we want to hit.

Call the bullseye L. We want his shots to get *closer and closer* to the bullseye. We want him to be able to *eventually* shoot an arrow within distance 1 of the target:

$$L - 1 \qquad L \qquad L + 1$$
$$a_{17}$$

And after waiting even longer, within distance $\dfrac{1}{2}$ of the target:

$$L - \tfrac{1}{2} \qquad L \qquad L + \tfrac{1}{2}$$
$$a_{92}$$

And after waiting still longer, within distance $\dfrac{1}{10}$ from the target:

$$L - \tfrac{1}{10} \quad L \quad L + \tfrac{1}{10}$$
$$a_{1792}$$

In fact, for *any desired accuracy* desired distance ϵ, we want one of his shots to eventually land within ϵ distance of the target:

$$L - \epsilon \, L \, L + \epsilon$$
$$a_{94305}$$

Intuitively, this means that we can *eventually* find a shot that is as *accurate* to the target as we want. Mathematically,

for any $\epsilon > 0$, there exists some integer index N such that $|a_N - L| < \epsilon$.

---

[1]Or for the young folks, that little girl from Brave.

Is this the definition of limit? *Almost.* But we have to *require even more.* Consider the previous example of

$$a_n = (-1)^n.$$

It is indeed true that for any $\epsilon > 0$ there exists some N such that

$$|a_N - 1| < \epsilon.$$

In fact, you are always guaranteed to exactly hit the target 1 on every even numbered shot. But intuitively, we feel the shots do not converge to 1.

We make one revision. Instead of having the archer eventually land a shot within distance ϵ of the target, we tyrannically demand that

*Eventually, **all shots** land within distance ϵ of the target.*

So like checking in at the Hotel California, eventually his shots *never leave.*

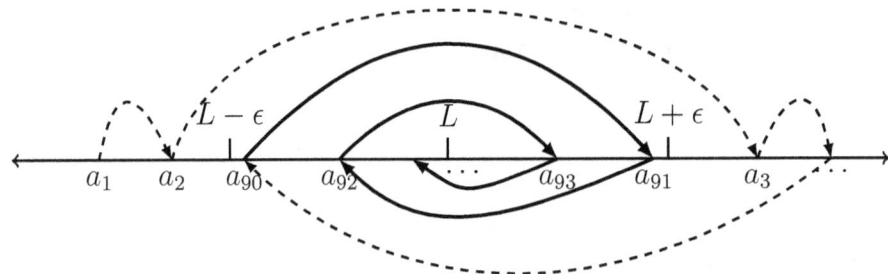

In this diagram, after a_{90}, all subsequent terms are trapped in the vortex between $L - \epsilon$ and $L + \epsilon$.

We now formalize this intuition with the following definition:

Definition. *We say L is the[1] **limit** of the sequence (a_n) if for any $\epsilon > 0$ there exists an integer N such that for all $i \geq N$*

$$|a_i - L| < \epsilon$$

We write

$$a_n \to L$$

to mean

The sequence (a_n) converges to L

We can alternatively write

$$\lim_{n \to \infty} a_n = L$$

which has the exact same meaning, but is read as

---

[1]In order to say that L is *the* limit rather than *a* limit, we need to prove that there can be only one limit!

The limit of the sequence (a_n) equals L

Even though the latter is more aesthetically pleasing, I tend to use the former notation because I prefer howit is read. Moreover, it reminds us that we need to *prove* that the limit is unique whereas the second notation already *asserts* this. I shall prove uniqueness of limits at the end of this lecture.

Note that the definition of a limit involves three magic words:

for **any** $\epsilon > 0$ there **exists** an N such that for **all** $i \geq N$

In order to prove that a sequence converges to some limit, we need to understand how to prove statements involving multiple instances of *any* and *exists*.

9.3 Proving Statements involving Multiple Quantifiers

*There **exists** a castle on a cloud,*
where I like to go when I'm asleep.
***Any** floor there I don't need to sweep.*
Here in my castle on a cloud.

-(Left)**Cose**tte

One of the reasons why students struggle with limit proofs is that they haven't been taught the mathematical *birds and bees*. And by the *birds and the bees*, I mean the *A's and the E's*:

Any and **Exists**

The logicians call these *logical quantifiers*. To sum up what I've taught you so far, typically:

- To prove a *universal (any)* statement, take an arbitrary element from the set, and show that it has the desired property.

- To prove an *existential (exists)* statement, construct a specific example.

Of course, there are some other ways to prove universal and existential statements. We've already given examples that use proof by cases, contradiction, and even mathematical axioms. But for the most part, we will use the two aforementioned techniques.

Some of our proofs have (innately) combined these two quantifiers. Particularly, we have worked through two combination types:

Any-Exists and **Exists-Any**

To prove these theorems, we only needed to combine the techniques above (in the right order[1]):

- To prove an *any-exists* statement, take an **arbitrary** element x_1 and **using** this x_1, **construct** some element x_2.

---

[1]The order of quantifiers does matter. You will see this when discussing "continuous functions" versus "uniformly continuous functions."

- To prove an *exists-any* statement, **find** some element x_1 such that an **arbitrary** element x_2 satisfies some property **involving** x_1.

Here is an example of proving an *any-exists* statement:

Example. *For **any** natural number n, there **exists** a perfect square greater than n.*

Proof: Let n be a natural number. **Using** n, we construct a number bigger than n that is a perfect square. Take

$$(n+1)^2$$

It is a perfect square and

$$(n+1)^2 = n^2 + (2n+1) > n^2 \geq n \qquad \square$$

Here, the arbitrary element n is used to construct the number we need, namely $(n+1)^2$.

As an example of an *exists-any* proof, we can show there exists a matrix (different from the identity) that does not change vector length:

Example. *There **exists** a non-identity matrix A such that for **any** $\vec{x} \in \mathbb{R}^2$*

$$\|A\vec{x}\| = \|\vec{x}\|$$

Proof: First we find a matrix. Intuitively, we know rotation shouldn't change vector length, so **choose**

$$A = \begin{bmatrix} \cos\theta & -\sin\theta \\ \sin\theta & \cos\theta \end{bmatrix}$$

Now we have to show that for any $\vec{x} \in \mathbb{R}^2$

$$\|A\vec{x}\| = \|\vec{x}\|$$

Let $\vec{x} \in \mathbb{R}^2$ be arbitrary. Then,

$$A\vec{x} = \begin{bmatrix} x_1 \cos\theta - x_2 \sin\theta \\ x_1 \sin\theta + x_2 \cos\theta \end{bmatrix}$$

Directly calculating the norm,

$$
\begin{aligned}
\|A\vec{x}\| &= \sqrt{\left(x_1\cos\theta - x_2\sin\theta\right)^2 + \left(x_1\sin\theta + x_2\cos\theta\right)^2} \\
&= \sqrt{\left(x_1\cos\theta\right)^2 + \left(x_1\sin\theta\right)^2 + \left(x_2\sin\theta\right)^2 + \left(x_2\cos\theta\right)^2} \\
&= \sqrt{x_1^2\left(\cos^2\theta + \sin^2\theta\right) + x_2^2\left(\sin^2\theta + \cos^2\theta\right)} \\
&= \sqrt{x_1^2 + x_2^2}
\end{aligned}
$$

In conclusion,
$$\|A\vec{x}\| = \|\vec{x}\|. \qquad \square$$

Here, we first constructed an object A and then showed that *any* vector $\vec{x}$ has its length unchanged by our *chosen* matrix A.

Before we move forward, I would like to re-emphasize that the second variable *depends* on the first.

- In the *any-exists* example, our constructed object was a *function* of the arbitrary n.

- In the *exists-any* example, our proof that any vector had its length unchanged by A relied on our *choice* of A.

In our limit proofs, N will be a *function* of ϵ.

Now, we are ready to deal with *any-exists-any* proofs. This is just a combination of the *any-exists* and *any* proofs:

> To prove an ***any-exists-any*** statement, take an arbitrary element x_1 and use it to **construct** an x_2, such that for an **arbitrary element** x_3, some property holds.

For example, here is a variation of the infinitude of primes proof:

Example. *For **any** finite set of prime numbers P, there **exists** a natural number n, such that for **any** prime in P, that prime does not divide n.*

Proof: Let P be a finite set of primes
$$P = \{p_1, p_2, \ldots, p_s\}.$$

We *use* these p_i to *construct* the n we need. Define
$$n = p_1 p_2 \ldots p_s + 1$$

Notice that for *any* p_i in P, dividing n by p_i always leaves remainder 1.
$$n = \underbrace{(p_1 p_2 \ldots p_{i-1} p_{i+1} \ldots p_s)}_{q} \underbrace{p_i}_{b} + \underbrace{1}_{r}.$$

Thus, for any prime $p_i \in P$, p_i does not divide n. $\qquad \square$

We started with some *arbitary* set of primes P and used it to *construct* a number n. Then for an arbitrary p_i in P, we showed that p_i did not divide n.

Once you feel comfortable with the preceding examples, you can try tackling limit proofs.

9.4   How to Prove a Sequence Converges to Some Limit

Start every limit proof with "Let ϵ be bigger than 0."

-Maksim Maydanskiy

Every (basic) limit proof should follow the same formula:

1. Let $\epsilon > 0$ i.e. fix ϵ as some positive number.

2. Rewrite the $|a_i - L| < \epsilon$ into an equivalent condition on i.

3. Choose an N such that all $i \geq N$, a_i satisfies the aforementioned condition.

Make sure that your choice of N is a function[1] of ϵ.

Let's start with a classic example:

Example. *The sequence (a_n) where*

$$a_n = \frac{1}{n}$$

converges to 0:

$$a_n \to 0.$$

Proof: Let $\epsilon > 0$. We need to find a corresponding N such that we can guarantee, for all $i \geq N$,

$$\left| \frac{1}{i} - 0 \right| < \epsilon$$

or simply

$$\frac{1}{i} < \epsilon$$

First, rewrite this condition to isolate i:

$$\frac{1}{\epsilon} < i \tag{$\star$}$$

Now, we have to *choose* an N such that for all $i \geq N$, condition $(\star)$ holds. Therefore, choose N to be an integer *strictly greater* than $\frac{1}{\epsilon}$. For example, choose

$$N = \left\lceil \frac{1}{\epsilon} \right\rceil + 1.$$

Then, for any $i \geq N$,

$$\frac{1}{\epsilon} \leq N \leq i$$

so the condition $(\star)$ is satisfied. Thus, we can conclude

$$a_n \to 0. \qquad \square$$

---

[1]This is literally one of my greatest pet-peeves. If you magically find an N that is independent of ϵ, you are either dealing with constants or mental health issues.

Note that it is **not enough** to choose

$$N = \left\lceil \frac{1}{\epsilon} \right\rceil.$$

This is because for some choices of ϵ (e.g. $\epsilon = \frac{1}{2}$) we have

$$\frac{1}{\epsilon} = N.$$

However, we require

$$\frac{1}{\epsilon} < N.$$

The next example involves negative numbers:

Example. *Define*

$$s_n = \frac{\cos(n)}{\sqrt{n}}$$

Then,

$$s_n \to 0.$$

Proof: Let $\epsilon > 0$. We need to find a corresponding N such that we can guarantee, for all $i \geq N$,

$$\left| \frac{\cos(i)}{\sqrt{i}} \right| < \epsilon$$

Because

$$\left| \frac{\cos(i)}{\sqrt{i}} \right| = \frac{|\cos(i)|}{|\sqrt{i}|} = \frac{|\cos(i)|}{\sqrt{i}},$$

this condition is equivalent to:

$$\frac{|\cos(i)|}{\sqrt{i}} < \epsilon.$$

To guarantee that the left hand side is less than ϵ, it is a lot easier to show that the *upper bound* of this quantity is less than ϵ. Since

$$|\cos(i)| \leq 1,$$

we have

$$\frac{|\cos(i)|}{\sqrt{i}} \leq \frac{1}{\sqrt{i}}.$$

Thus it suffices to prove

$$\frac{1}{\sqrt{i}} < \epsilon.$$

Isolating i, we can rewrite this condition as

$$\frac{1}{\epsilon^2} < i.$$

Choose
$$N = \left[\frac{1}{\epsilon^2}\right] + 1$$

Then, for any $i \geq N$,
$$\frac{1}{\epsilon^2} < N \leq i$$

as needed. Thus, we can conclude
$$s_n \to 0. \qquad\qquad \square$$

Here's a harder example:

Example. *Define*
$$a_n = \frac{2^n}{n!}.$$

Then,
$$a_n \to 0.$$

Proof: Let $\epsilon > 0$. Then, we need to find an N such that for all $i \geq N$,
$$\frac{2^i}{i!} < \epsilon$$

Again, it is easier to find a *upper bound* for the (LHS) and then show that this upper bound is less than ϵ.

Expand the (LHS):
$$\frac{2^i}{i!} = \frac{\overbrace{2 \cdot 2 \cdot 2 \cdot \ldots \cdot 2}^{i \text{ times}}}{i \cdot (i-1) \cdot (i-2) \cdot \ldots \cdot 3 \cdot 2 \cdot 1}$$

and rewrite this as a product of fractions:
$$\frac{2}{i} \cdot \frac{2}{i-1} \cdot \ldots \cdot \frac{2}{4} \cdot \frac{2}{3} \cdot \frac{2}{2} \cdot \frac{2}{1}$$

Notice that the following terms are *at most 1*:
$$\frac{2}{i} \cdot \underbrace{\frac{2}{i-1}}_{\leq 1} \cdot \ldots \cdot \underbrace{\frac{2}{4}}_{\leq 1} \cdot \underbrace{\frac{2}{3}}_{\leq 1} \cdot \underbrace{\frac{2}{2}}_{\leq 1} \cdot \frac{2}{1}$$

Therefore, this product is bounded above by
$$\frac{2}{i} \cdot 1 \cdot 1 \cdot 1 \cdot \ldots \cdot 1 \cdot 2.$$

giving us
$$\frac{2^i}{i!} \leq \frac{4}{i}.$$

Now we need only show that the upper bound is less than ϵ:

$$\frac{4}{i} < \epsilon$$

But this is equivalent to showing

$$\frac{4}{\epsilon} < i.$$

Therefore, choose

$$N = \left\lceil \frac{4}{\epsilon} \right\rceil$$

Then, for any $i \geq N$,

$$\frac{4}{\epsilon} < N \leq i$$

as needed. In conclusion,

$$a_n \to 0. \qquad \square$$

In the preceding proof, notice that we did not to use the fact that

$$\frac{2}{i} \leq 1.$$

If we did, our bound would be *too big*:

$$\frac{2^i}{i!} \leq 2.$$

No matter how big we require i to be, we cannot manipulate 2 to be less than ϵ. However, by leaving $\frac{2}{i}$ alone, we got

$$\frac{2^i}{i!} \leq \frac{4}{i}$$

in which we can indeed find a condition on i to ensure

$$\frac{4}{i} < \epsilon.$$

The next example involves a non-zero limit:

Example. *Define*

$$a_n = \frac{n-1}{n}$$

Then,

$$a_n \to 1.$$

Proof: Let $\epsilon > 0$. We want to find an N such that for all $i \geq N$,

$$\left| \frac{i-1}{i} - 1 \right| < \epsilon$$

Since

$$\frac{i-1}{i} - 1 = \frac{i-1}{i} - \frac{i}{i}$$

$$= \frac{-1}{i}$$

we can simplify this condition to

$$\left| \frac{-1}{i} \right| < \epsilon.$$

Moreover, we can simplify the absolute value

$$\frac{1}{i} < \epsilon$$

and isolate i:

$$\frac{1}{\epsilon} < i.$$

Again, choose

$$N = \left\lceil \frac{1}{\epsilon} \right\rceil + 1$$

Then, for any $i \geq N$,

$$\frac{1}{\epsilon} < N \leq i$$

Thus, we can conclude

$$a_n \to 1 \qquad \square$$

Here's an example you saw when studying horizontal asymptotes:

Example. *Define*

$$b_n = \frac{3n^2 + 8}{2n^2 + 4}$$

Then,

$$b_n \to \frac{3}{2}.$$

Proof: Let $\epsilon > 0$. We want to find an N such that for all $i \geq N$,

$$\left| \frac{3i^2 + 8}{2i^2 + 4} - \frac{3}{2} \right| < \epsilon.$$

Since

$$\frac{3i^2 + 8}{2i^2 + 4} - \frac{3}{2} = \frac{3i^2 + 8}{2i^2 + 4} - \frac{3i^2 + 6}{2i^2 + 4}$$

$$= \frac{2}{2i^2 + 4}$$

$$= \frac{1}{i^2 + 2}$$

we need to find an N such that

$$\left| \frac{1}{i^2 + 2} \right| < \epsilon.$$

Dropping the absolute value,

$$\frac{1}{i^2 + 2} < \epsilon$$

and isolating i, we get[1]

$$\sqrt{\frac{1}{\epsilon} - 2} < i$$

Choose

$$N = \left\lceil \sqrt{\frac{1}{\epsilon} - 2} \right\rceil + 1$$

Then, for an arbitrary $i \geq N$,

$$\sqrt{\frac{1}{\epsilon} - 2} < N \leq i$$

Thus, we conclude

$$b_n \to \frac{3}{2} \qquad \qquad \square$$

9.5   Limit Properties: Addition and Scaling

Oftentimes, we do not want to refer to the original $\epsilon - N$ definition of limit. Instead, we use *limit properties*. There are two main reasons:

- **The limit may be intuitively obvious.**
  For example, we can feel it in our bones that the limit of

  $$a_n = 3 \cdot \frac{n-1}{n}$$

  is 3. Why? Because we already know the limit of

  $$b_n = \frac{n-1}{n}$$

  is 1. Intuitvely,

  > *Scaling a convergent sequence by a constant scales the limit by the same constant.*

  As another example, the limit of

  $$x_n = \frac{n-1}{n} + \frac{2^n}{n!}$$

  should be 1. This is because the limits of the individual sequences

  $$y_n = \frac{n-1}{n} \qquad z_n = \frac{2^n}{n!}$$

  sum to 1, and intuitively,

---

[1]Technically, we should make a separate case for $\epsilon > .5$. However, for pedagogical reasons, ignore this technicality.

The limit of a sum of convergent sequences is the sum of their individual limits.

- **We can break the terms of a sequence into easier parts.**
  Suppose we wanted to compute the limit of

$$a_n = \frac{(n-1)^{10}}{n^{10}}$$

The *bone-headed* thing to do would be to expand $(n-1)^{10}$ as an ugly polynomial. Instead, we can show that

A product of convergent sequences converges to the product of their individual limits.

In particular

$$a_n = \frac{(n-1)}{n} \cdot \frac{(n-1)}{n} \cdot \ldots \cdot \frac{(n-1)}{n}$$

where each $\frac{(n-1)}{n}$ converges to 1. So we conclude:

$$a_n \to 1.$$

- **Proving the limit directly from the definition can be too difficult.**
  For example, suppose you are given the sequence

$$a_n = \sqrt[n]{n}$$

and you want to prove

$$a_n \to 1$$

No, we can't just apply L'Hospital's rule. You don't know how to prove it (nor do 99% of high school students). If you tried direct expansion, you would get stuck showing

$$n < (1+\epsilon)^n$$

Instead, we can apply a more advanced limit property known as the *Sandwich Theorem*,[1] which is proven in the next lecture.

We begin with proving the most basic limit property, scaling:

Theorem. *Let*

$$a_n \to L$$

and let k be a constant. Then, the sequence formed from scaling each term by k

$$(ka_n)$$

converges and

$$ka_n \to kL.$$

---

[1]You could argue that we can still write everything directly in terms of the ϵ-N since the Sandwich Theorem follows from the same definition. But it's like applying a function in Java. We can either use 5 lines of code where each line represents 100 lines of code, or plug in 500 lines of code. Alternatively you can think of this as our matrix multiplication example: would you rather write out all 1000+ entries or a single arbitrary entry?

Proof Summary:

1. Let $\epsilon > 0$.

2. Use convergence of (a_n) with choice $\frac{\epsilon}{k}$ to get corresponding N'.

3. Choose $N = N'$.

Proof: Assume $k \neq 0$ (since the zero sequence obviously converges). Let $\epsilon > 0$. We want to find an N such that for all $i \geq N$

$$|ka_i - kL| < \epsilon$$

By absolute value properties, we can pull out the k:

$$|k||a_i - L| < \epsilon$$

and thus rewrite our condition as

$$|a_i - L| < \frac{\epsilon}{k}.$$

Since we are given

$$a_n \to L,$$

for any $\epsilon' > 0$ we can a find a corresponding N' such that for all $j \geq N'$,

$$|a_j - L| < \epsilon'.$$

Choose

$$\epsilon' = \frac{\epsilon}{k}.$$

Then there is a corresponding N' such that for all $j \geq N'$,

$$|a_j - L| < \epsilon' = \frac{\epsilon}{k}$$

But lo and behold, this is *exactly* the condition we wanted to show. Just let $N = N'$ and we're done!□

Notice, that we had to add $'$ and change i to j. If we kept the same variables, we would have completely changed the *meaning* of the statement! Generally,

---

Math Mantra: Don't be a dummy with dummy variables!

Mull over the dummy variables in the proof and think about the fact that we can choose an ϵ' and corresponding N'. When you are comfortable enough to explain the proof to your grand pappy over the phone, move on to the sum property:

Theorem. *Let*

$$a_n \rightarrow L_1$$
$$b_n \rightarrow L_2$$

Then, the sequence formed by adding each term point-wise

$$(a_n + b_n)$$

converges and

$$(a_n + b_n) \to L_1 + L_2.$$

Proof Summary:

1. Let $\epsilon > 0$.

2. Apply the convergence definition of a_n with choice $\epsilon_1 = \frac{\epsilon}{2}$ to get corresponding N_1.

3. Apply the convergence definition of b_n with choice $\epsilon_2 = \frac{\epsilon}{2}$ to get corresponding N_2.

4. Choose $N = \max\{N_1, N_2\}$.

Proof: Let $\epsilon > 0$. We want to find an N such that

$$|a_i + b_i - L_1 - L_2| < \epsilon.$$

for all $i \geq N$. Rearranging terms,

$$|(a_i - L_1) + (b_i - L_2)| < \epsilon$$

Again, it is often easier to show that some *upper bound* for the (LHS) is less than ϵ. Applying the triangle inequality on the (LHS),

$$|(a_i - L_1) + (b_i - L_2)| \leq |a_i - L_1| + |b_i - L_2|$$

Therefore, if we can find an N such that all $i \geq N$,

$$|a_i - L_1| + |b_i - L_2| < \epsilon,$$

we are done!

By the definition of

$$a_n \to L_1,$$

we know that for any $\epsilon_1 > 0$, we can find an N_1 such that for all $i \geq N_1$,

$$|a_i - L_1| < \epsilon_1.$$

In particular, if we choose $\epsilon_1 = \frac{\epsilon}{2}$, there is an N_1 such that for all $i \geq N_1$,

$$|a_i - L_1| < \frac{\epsilon}{2}.$$

Likewise, using the definition of

$$b_n \to L_2,$$

with the choice $\epsilon_2 = \frac{\epsilon}{2}$, we can find an N_2 such that for all $i \geq N_2$,

$$|b_i - L_2| < \epsilon_2$$

Combining these N-hypotheses, if

$$i \geq N_1 \text{ and } i \geq N_2$$

then in fact

$$\underbrace{\left|a_i - L_1\right|}_{<\frac{\epsilon}{2}} + \underbrace{\left|b_i - L_2\right|}_{<\frac{\epsilon}{2}} < \epsilon$$

Therefore, to ensure that *both* hypotheses are satisfied, choose N to be the **bigger** of N_1 and N_2:

$$N = \max\{N_1, N_2\}.$$

Therefore, when $i \geq N$, we know $i \geq N_1$ and $i \geq N_2$, so

$$\underbrace{\left|a_i - L_1\right|}_{<\frac{\epsilon}{2}} + \underbrace{\left|b_i - L_2\right|}_{<\frac{\epsilon}{2}} < \epsilon.$$

In conclusion,

$$(a_n + b_n) \to L_1 + L_2. \qquad \square$$

9.6   Limit Properties: Product

We also have a product rule for convergent sequences. However, unlike scaling and summing, this proof is not straightforward. Particularly, we need two things:

- An algebraic trick

- A nice idea

The algebraic trick is the same one we have used ever since we completed the square in Algebra II: just add 0 (you'll see)!

As for the nice idea, we need to prove

Convergent sequences are bounded.

Why do we know this is the idea that we are going to need?

```
Math Mantra:  Nice ideas usually come from reaching an impasse in the proof.
         When you get stuck, prove a new result to help you out.
```

To show a sequence is bounded *above* by a constant, just fix ϵ to be some number, say $\epsilon = 1$. Then we know that the "tail end of the sequence" is bounded above by $L + 1$:

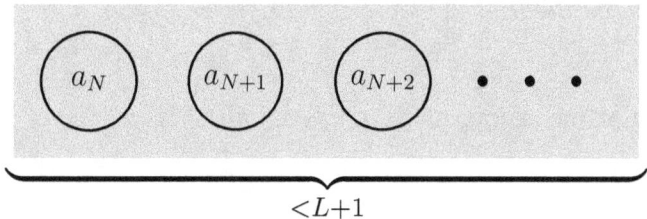

Moreover, the *finitely* many terms at the beginning of the sequence are bounded by the greatest among them:

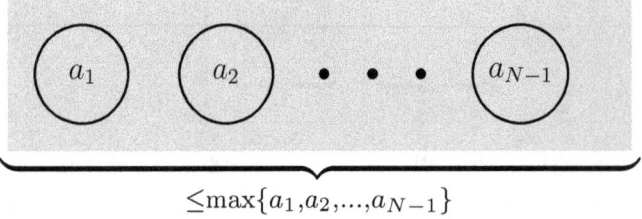

$$\leq \max\{a_1, a_2, ..., a_{N-1}\}$$

Therefore, the entire sequence is bounded above by

$$K_1 = \max\{L+1, a_1, a_2, \ldots, a_{N-1}\}$$

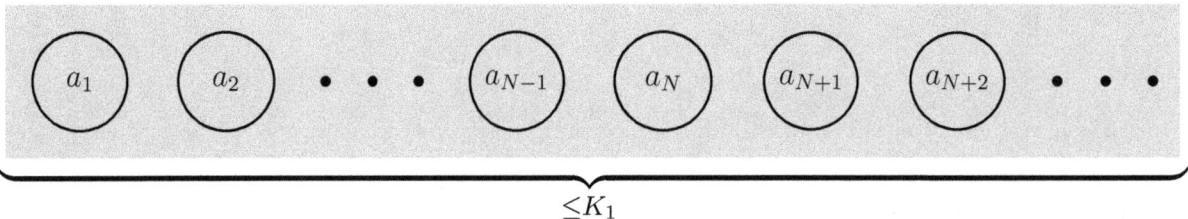

$$\leq K_1$$

A similar process is used to prove that the sequence is bounded below. Now we fill the details to get a formal proof:

Theorem. *Convergent sequences are bounded. In other words, if*

$$a_n \to L$$

then there exists some $K > 0$ such that

$$|a_i| \leq K$$

for all $i = 1, 2, \ldots$.

Proof Summary:

- We prove the sequence is bounded above by some K_1:
    - Set $\epsilon = 1$.
    - K_1 is the biggest value among $a_1, a_2, \ldots a_{N-1}, L+1$.
- We prove the sequence is bounded below by some K_2:
    - Set $\epsilon = 1$.
    - K_2 is the smallest value among $a_1, a_2, \ldots a_{N-1}, L-1$.
- Let $K = \max\{|K_1|, |K_2|\}$

Proof: First, we prove that the convergent sequence is **bounded above**, i.e. there exists K_1 such that

$$a_i \leq K_1$$

for all i. Apply the convergence definition with the choice $\epsilon = 1$. Then we know there is an N such that for all $i \geq N$,

$$|a_i - L| < 1.$$

or equivalently,

$$L - 1 < a_i < L + 1.$$

In particular, past index N, a_i cannot be greater than $L + 1$. So in fact,

$$a_i < L + 1$$

for all $i \geq N$. Hence, there are only *finitely* many terms that are not automatically bounded above by $L + 1$, namely:

$$a_1, a_2, \ldots, a_{N-1}.$$

Therefore, *every* term in the entire sequence is bounded above by

$$K_1 = \max\{a_1, a_2, \ldots, a_{N-1}, L + 1\}$$

To prove that the sequence is **bounded below**, we follow a similar argument: we want to find K_2 such that

$$K_2 \leq a_i$$

for all i. Choose $\epsilon = 1$. By the definition of convergence, we know that there is an N such that

$$|a_i - L| < 1$$

for all $i \geq N$. In particular,

$$L - 1 < a_i$$

for all $i \geq N$. Therefore, the entire sequence is bounded below by

$$K_2 = \min\{a_1, a_2, \ldots, a_{N-1}, L - 1\}$$

Since our sequence is bounded above and below,

$$K_2 \leq a_i \leq K_1$$

for all i. Define

$$K = \max\{|K_1|, |K_2|\}.$$

Then for all i, we have

$$-K \leq K_2 \leq a_i \leq K_1 \leq K.$$

Hence,

$$|a_i| \leq K$$

for all i. □

Now that we have our nice idea, we can prove the product property:

Theorem. *If*

$$a_n \;\to\; L_1$$
$$b_n \;\to\; L_2$$

Then the sequence formed taking the point-wise product

$$(a_n b_n)$$

converges and

$$a_n b_n \to L_1 L_2$$

Proof Summary:

- In the ϵ-condition, add $\underbrace{a_i L_2 - a_i L_2}_{=0}$.

- Use the triangle inequality and the boundedness of convergent sequences to establish an upper bound.

- Apply the convergence of a_n with choice $\dfrac{\epsilon}{2|L_2|}$ to get corresponding N_1.

- Apply the convergence of b_n with choice $\dfrac{\epsilon}{2K}$ to get corresponding N_2.

- Choose $N = \max\{N_1, N_2\}$.

Proof: Let $\epsilon > 0$. We want to show that there exists an N such that for all $i \geq N$,

$$|a_i b_i - L_1 L_2| < \epsilon.$$

Using the awesome trick of adding 0, rewrite the ϵ-condition as:

$$|a_i b_i \underbrace{-a_i L_2 + a_i L_2}_{=0} - L_1 L_2| < \epsilon$$

Equivalently,

$$|(a_i b_i - a_i L_2) + (a_i L_2 - L_1 L_2)| < \epsilon$$

Therefore, it suffices to show that some *upper bound* for

$$|(a_i b_i - a_i L_2) + (a_i L_2 - L_1 L_2)|$$

is less than ϵ. Using our trusty triangle inequality,

$$|(a_i b_i - a_i L_2) + (a_i L_2 - L_1 L_2)| \leq |a_i b_i - a_i L_2| + |a_i L_2 - L_1 L_2|.$$

By absolute value properties, the (RHS) is

$$|a_i||b_i - L_2| + |L_2||a_i - L_1|.$$

Now we apply our nice idea. Since, any convergent sequence is bounded, there exists a $K > 0$ such that for any i,

$$|a_i| \leq K.$$

Therefore,

$$\underbrace{|a_i|}\, |b_i - L_2| + |L_2||a_i - L_1| \leq \underbrace{K}\, |b_i - L_2| + |L_2||a_i - L_1|$$

Since

$$a_i \to L_1$$

we know there[1] is some N_1 such that for every $i \geq N_1$,

$$|a_i - L_1| < \frac{\epsilon}{2|L_2|}$$

Likewise, we know there exists an N_2 such that for every $i \geq N_2$

$$|b_i - L_2| < \frac{\epsilon}{2K}$$

To ensure that both ϵ-conclusions hold, let

$$N = \max(N_1, N_2).$$

Then, for $i \geq N$,

$$K \underbrace{|b_i - L_2|}_{< \frac{\epsilon}{2K}} + |L_2| \underbrace{|a_i - L_1|}_{< \frac{\epsilon}{2|L_2|}} < \epsilon. \qquad \square$$

9.7 Uniqueness of Limits

We conclude our first lecture on limits with one more fundamental fact. First, let's take another look at our original definition of *limit*:

Definition. *We say L is **the** limit of the sequence (a_n) if for any $\epsilon > 0$ there exists an integer N such that for all $i \geq N$*

$$|a_i - L| < \epsilon$$

Notice that I wrote '*the* limit and not *a* limit. By using the word *the*, I imply that the limit is *unique*. It's like saying

Chuck Norris is <u>the</u> king of the world.

versus

Chuck Norris is <u>a</u> king of the world.

---

[1]Careful! At this point I divide by $|L_2|$. Therefore, we will need to consider the case $L_2 = 0$ separately. This proof is an easy exercise and I leave it to the enthusiastic reader to complete.

Why am I making such a big deal out of a single word?

- We don't want multiple limits! Particularly,

$$a_n = (-1)^n$$

  should not converge to both -1 and 1.

- We will use uniqueness to *solve* for unknown limits. For example, if we can show

$$a_n \rightarrow 3x+1$$
$$a_n \rightarrow 2x+5$$

  then by uniqueness, we know that

$$3x + 1 = 2x + 5.$$

  Thus $x = 4$ and therefore

$$a_n \rightarrow 13.$$

Let's prove that the limit of a sequence is unique:

Theorem. *If*

$$a_n \rightarrow L_1$$
$$a_n \rightarrow L_2$$

then

$$L_1 = L_2.$$

Proof Summary:

- If we can prove $|L_1 - L_2| < \epsilon$ for all $\epsilon > 0$, then it must be the case $L_1 = L_2$.

- Let $\epsilon > 0$.

- Add 0: $|L_1 - L_2| = |L_1 - \underbrace{a_i + a_i}_{0} - L_2|$

- Use triangle inequality and limit definition to show the upper bound is less than ϵ.

Proof: Note that if

$$|L_1 - L_2| < \epsilon$$

for any $\epsilon > 0$, then $L_1 - L_2$ **must be** 0. Otherwise, the choice

$$\epsilon = |L_1 - L_2|$$

forces

$$|L_1 - L_2| < \underbrace{|L_1 - L_2|}_{\epsilon}$$

which is absurd!

Therefore, if we can prove for any $\epsilon > 0$,

$$|L_1 - L_2| < \epsilon,$$

we can conclude

$$L_1 = L_2.$$

Let $\epsilon > 0$. Rewrite $|L_1 - L_2|$ by adding zero

$$|L_1 \underbrace{-a_i + a_i}_{0} - L_2|$$

and regrouping

$$|(L_1 - a_i) + (a_i - L_2)|.$$

Applying triangle inequality,

$$|(L_1 - a_i) + (a_i - L_2)| \leq |L_1 - a_i| + |a_i - L_2|.$$

Now we just have to show this *upper bound* is less than ϵ. Since,

$$a_n \to L_1$$

we know for the particular choice of $\dfrac{\epsilon}{2}$, we can find an N_1 such that for all $i \geq N_1$

$$|a_i - L_1| < \frac{\epsilon}{2}.$$

Likewise, since

$$a_n \to L_2$$

we can find an N_2 such that for all $i \geq N_2$

$$|a_i - L_2| < \frac{\epsilon}{2}$$

Let $N = \max\{N_1, N_2\}$. Then for all $i \geq N$,

$$\underbrace{|L_1 - a_i|}_{< \frac{\epsilon}{2}} + \underbrace{|a_i - L_2|}_{< \frac{\epsilon}{2}} < \epsilon$$

Therefore

$$|L_1 - L_2| < \epsilon$$

and

$$L_1 = L_2. \qquad \square$$

New Notation

Symbol	Reading	Example	Example Translation
(a_n)	The sequence with each term a_n.	$\left(\dfrac{1}{n}\right)$ converges to 0.	The sequence whose n-th term is $\dfrac{1}{n}$ converges to 0.
$a_n \to L$	The sequence (a_n) converges to L.	$\dfrac{1}{n} \to 0$	The sequence $\left(\dfrac{1}{n}\right)$ converges to 0.
$\lim\limits_{n\to\infty} a_n = L$	The limit of sequence (a_n) equals L.	$\lim\limits_{n\to\infty} \sqrt[n]{n} = 1$	The limit of the sequence $(\sqrt[n]{n})$ is 1.
ϵ	Epsilon	Otis calls his students ϵ's.	Otis calls his students epsilons.

Lecture 10

Being Bolzy

Come on you math majors if you want to be free,
From Corporate America you listen to me.
You've got a sequence that you built from your approximating tweakins,
And you really need to find a convergent subsequence,
So you ask my man Bolzano and his homie Weierstrass,
Who've found you a solution with a trick that's really boss.

- Stephen Sawin

Goals: Today, we prove the infamous Bolzano-Weierstrass theorem. But in order to do so, we first need to prove the Monotone Convergence Property and the Sandwich Theorem. These three theorems will be useful tools in proving limit properties.

10.1 The Next Big Thing

This is the biggest theorem since sliced bread. Or rather, the biggest theorem since Cauchy-Schwarz. And like Cauchy-Schwarz, Bolzano-Weierstrass has multiple proofs, so it must be super important. In fact, there is even a Bolzano-Weierstrass Rap (YouTube it)!

But in order to understand Bolzano-Weierstrass, we must first understand *subsequences*. A subsequence is just a sequence formed by picking out terms of a larger sequence. For example, given the sequence (a_n):

$$1, \quad 4, \quad 9, \quad 16, \quad 25, \quad 36, \quad 49, \quad 64, \quad \dots$$

we can pick out specific terms

$$1, \quad \boxed{4}, \quad \boxed{9}, \quad 16, \quad \boxed{25}, \quad 36, \quad \boxed{49}, \quad 64, \quad \dots$$

to form the subsequence

$$
\begin{aligned}
a_{n_1} &= 4 \\
a_{n_2} &= 9 \\
a_{n_3} &= 25 \\
a_{n_4} &= 49 \\
&\vdots
\end{aligned}
$$

Formally,

Definition. *Given an original sequence* $(a_n)_{n=1}^{\infty}$, *a **subsequence** is a sequence*

$$(a_{n_j})_{j=1}^{\infty}$$

where (n_i) *is an increasing sequence of indices:*

$$1 \leq n_1 < n_2 < n_3 \ldots$$

Note that:

- $(a_n)_{n=1}^{\infty}$ is a sequence indexed by n:

$$a_1, \quad a_2, \quad a_3, \quad a_4, \quad a_5, \quad \ldots$$

whereas $(a_{n_j})_{j=1}^{\infty}$ is a sequence indexed by j:

$$a_{n_1}, \quad a_{n_2}, \quad a_{n_3}, \quad a_{n_4}, \quad a_{n_5}, \quad \ldots$$

We will **continue to write** $(a_n)_{n=1}^{\infty}$ as (a_n) and $(a_{n_j})_{j=1}^{\infty}$ as (a_{n_j}) while assuming the indexing convention.

- The increasing n_i condition simply means

*Move along the sequence, picking each term **after** the previous one, without ever going back to an earlier term in the sequence.*

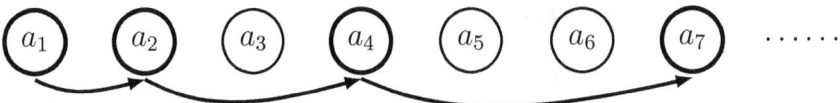

In our diagram,

$$\begin{aligned} n_1 &= 1 \\ n_2 &= 2 \\ n_3 &= 4 \\ n_4 &= 7 \\ &\vdots \end{aligned}$$

To get practice with this definition, let's prove the following *fundamental* result:

Every subsequence of a convergent sequence also converges.
Moreover, it converges to the same limit as the original sequence.

This is a very intuitive result:

Theorem. *If*

$$a_n \to L,$$

then any subsequence (a_{n_i}) *converges to the* **same limit***:*

$$a_{n_i} \to L$$

Proof: Let $\epsilon > 0$. Since (a_n) converges to L, there exists some N such that for all $i \geq N$,

$$|a_i - L| < \epsilon.$$

Observe that

$$n_i \geq i.$$

This is because (n_i) is an *increasing* sequence of positive integers; therefore, (n_i) increases at least as quickly as the slowest possible increasing sequence of positive integers $1, 2, 3, \ldots$.

By our observation, whenever $i \geq N$,

$$n_i \geq i \geq N,$$

so we have

$$|a_{n_i} - L| < \epsilon.$$

Therefore, for all $i \geq N$,

$$|a_{n_i} - L| < \epsilon. \qquad \square$$

Even though this was a very simple proof, do not underestimate its value! You'll exploit this result *numerous times*. Particularly,

Suppose you know that (a_n) *converges but* **you do not know the value of the limit** L.

Then you can *extract* the value of L by finding the limit of a *particular* subsequence!

Now that you understand subsequences, we can state the Bolzano-Weierstrass theorem:

Every bounded sequence has a convergent subsequence.

In some cases, it is obvious. For example, in a bounded sequence like

$$1, \quad 0, \quad 1, \quad 0, \quad 1, \quad 0, \quad 1, \quad 0, \quad \ldots$$

we can pick out all the even terms:

$$1, \quad \boxed{0}, \quad 1, \quad \boxed{0}, \quad 1, \quad \boxed{0}, \quad 1, \quad \boxed{0}, \quad \ldots$$

to get the convergent subsequence (a_{n_j}) of all zeros. Even in an sequence that lists all the rationals on $[0, 1]$, you could pick something like

$$\frac{0}{1}, \quad \boxed{\frac{1}{1}}, \quad \frac{0}{2}, \quad \boxed{\frac{1}{2}}, \quad \frac{2}{2}, \quad \frac{0}{3}, \quad \boxed{\frac{1}{3}}, \quad \frac{2}{3}, \quad \frac{3}{3}, \quad \frac{0}{4}, \quad \boxed{\frac{1}{4}}, \quad \frac{2}{4}, \quad \frac{3}{4}, \quad \frac{4}{4}, \quad \ldots$$

But what if Bart Simpson decided to dedicate his life to picking crazy numbers from $[-100, 100]$ one at a time:

$$\frac{\pi^2}{6}, \quad e^e, \quad \frac{\pi^2}{90}, \quad 2^{\sqrt{3}}, \quad \ldots$$

Here, constructing a convergent subsequence is *not obvious*.

The key is to use the fact the these numbers are *bounded* to plot them:

$-K$ K

From this picture, we are going to **devise a simple plan to construct a convergent subsequence.**

By the way, we never answered the question:

Why do we care about Bolzano-Weierstrass ?

Here's a great reason:

Almost all of the *magical* theorems in this course will stem from this seemingly innocuous fact.

Yes, I'm serious. And if you want a ghost of a chance of surviving this course, much like Indiana Jones you need to take a leap of faith before you can achieve the Holy Grail.

To prove Bolzano-Weierstrass we will need two major theorems:

- Monotone Convergence Property

- Sandwich Theorem

10.2   Monotone Convergence Property

Imagine Barbossa forcing Jack Sparrow to walk the plank. Jack can either

- Move forward slightly.

- Stay still.

However, he cannot move closer to the ship (at the risk of getting skewered) and can't jump into the ocean (at the risk of drowning). Therefore, Jack gets *arbitrarily close* to some fixed position.

The Monotone Convergence Property is the *same* idea: we have a sequence that either increases or stays the same (we call such a sequence *monotonically increasing*). It cannot backtrack, and it is bounded above. So eventually it gets *smooshed* up somewhere:

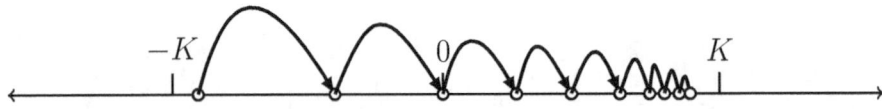

The same is true of a monotonically decreasing sequence. Formally we say

All bounded monotonic sequences converge.

We only prove the theorem for monotonically increasing sequences since the case of a monotonically decreasing sequence is almost verbatim.

Theorem (Monotone Convergence Property). *If*

$$a_i \leq a_j$$

for $i < j$ and there exists K such that

$$a_i \leq K$$

for all i, then (a_i) converges to some real number.

Proof Summary:

- Consider the *set* of sequence terms. By the *Completeness Axiom*, there is a least upper bound S.

- Show that the sequence converges to S:

  - Argue there is at least one element in the interval $(S - \epsilon, S]$.
  - All future elements remain in this interval.

Proof: Consider the *set* of sequence terms:

$$A = \{a_i \,|\, i \in \mathbb{N}\} \,.$$

Since this set is bounded and nonempty, by the Completeness Axiom, A has a least upper bound S. We now show (a_i) converges to S:

Let $\epsilon > 0$. Then we want to show that there exists N such that for all $i \geq N$,

$$|a_i - S| < \epsilon$$

First, there must be at least one a_q that lies in the interval

$$S - \epsilon < a_q \leq S.$$

Otherwise, $S - \epsilon$ would be an upper bound, contradicting that S is the least upper bound of A. Secondly, for all $i \geq q$

$$|a_i - S| < \epsilon$$

Suppose not. Then there must exists some $j > q$ such that either

$$a_j \leq S - \epsilon \quad \text{or} \quad a_j \geq S + \epsilon$$

In the case
$$a_j \leq S - \epsilon$$
we have
$$a_j \leq S - \epsilon < a_q$$
contradicting that (a_i) is monotonic increasing. The second case,
$$a_j \geq S + \epsilon$$
contradicts that S is an upper bound of the sequence.

Thus, for all $i \geq q$
$$|a_i - S| < \epsilon$$
Therefore, set $N = q$ and we're done!   □

At this point, *every* math textbook likes to point out that **this theorem relies heavily on the Completeness Axiom**. In fact, we can prove:

> *The Completeness Axiom is logically* **equivalent** *to the Monotone Convergence Property*

This means we can replace our Completeness Axiom by the Monotone Convergence Property in the original Field Axioms. But like choosing between Team Edward and Team Jacob, *it doesn't matter*.

I personally prefer the Monotone Convergence Property: to prove a sequence converges to a real number we just need to show the sequence is

- Bounded

- Monotonic

But showing these properties is pretty easy:

JUST USE INDUCTION!

Sequences are *meant* for induction, just like Starships are *meant* to fly. Don't believe me? Here's a fun example:

If you are a Khan Academy addict, you may have run into the magic number known as the Golden Ratio,
$$\frac{1 + \sqrt{5}}{2}$$
It tends to show up in expressions that involve *self-circularity*. For example, you may have seen the following (**incorrect**) proof:

Example.
$$\sqrt{1 + \sqrt{1 + \sqrt{1 + \ldots}}} = \frac{1 + \sqrt{5}}{2}$$

Bad Proof: Define

$$y = \sqrt{1 + \sqrt{1 + \sqrt{1 + \dots}}}$$

Then,

$$y^2 = 1 + \sqrt{1 + \sqrt{1 + \dots}}$$

But lo and behold, the right-hand side contains our original expression. Therefore,

$$y^2 = 1 + y$$

Using the quadratic equation and selecting the positive solution since y is clearly non-negative, we get

$$y = \frac{1 + \sqrt{5}}{2}.$$

This proof has *major* problems. The first is :

How do we know that we can give $\sqrt{1 + \sqrt{1 + \sqrt{1 + \dots}}}$ *a name?*

The name y suggests a finite numner, but how do we know this expression doesn't *explode?* If it did, then **bad things would happen**.

For example, consider the explosion

$$S = 1 + 1 + 1 + \dots$$

Exploiting circularity as before

$$S - 1 = S.$$

In other words,

$$1 = 0.$$

OUCH.

Secondly,

*Who said we can even take **infinite** summations and square roots?*

To quote Professor Simon,

Math is not a mystical study[1]!

We are really looking at tricky notation for a *limit*, namely the limit of the sequence defined by

$$a_1 = \sqrt{1}$$
$$a_{n+1} = \sqrt{1 + a_n} \quad \text{for all } n \geq 1$$

Here is a proper proof using the Monotone Convergence Property and limits:

---

[1]Despite what some logicians believe (*cough* transfinite induction *cough*).

Example. *Define a sequence by:*

$$a_1 = \sqrt{1}$$
$$a_{n+1} = \sqrt{1+a_n} \quad \text{for all } n \geq 1$$

Then,

$$a_n \to \frac{1+\sqrt{5}}{2}.$$

Proof: In **Lecture 4**, we already proved this sequence is bounded by 2. Therefore to show convergence, we need only prove (a_n) is increasing.

We use induction on n to prove the property

$$P_n : a_{n+1} - a_n \geq 0$$

- BASE CASE, $n = 1$
  Obvious since

$$\sqrt{1+\sqrt{1}} - 1 \geq 0$$

- INDUCTIVE STEP
  Let $k \geq 1$. Assume P_k is true:

$$a_{k+1} - a_k \geq 0.$$

We want to show property P_{k+1}:

$$a_{k+2} - a_{k+1} \geq 0.$$

By definition

$$a_{k+2} = \sqrt{1+a_{k+1}}$$
$$a_{k+1} = \sqrt{1+a_k}$$

Therefore, P_{k+1} is equivalent to

$$\underbrace{\sqrt{1+a_{k+1}}}_{a_{k+2}} - \underbrace{\sqrt{1+a_k}}_{a_{k+1}} \geq 0$$

i.e.

$$\sqrt{1+a_{k+1}} \geq \sqrt{1+a_k}.$$

But from the inductive hypothesis,

$$a_{k+1} \geq a_k.$$

Adding 1 to both sides and using the fact that square roots preserve non-negative inequalities (for the billionth time),

$$\sqrt{1+a_{k+1}} \geq \sqrt{1+a_k}.$$

as needed.

Thus (a_n) is increasing and bounded, so by the Monotone Convergence Property,

$$a_n \to L$$

for some L. Awesome, we have an actual *number* L to work with!

Now, consider the sequence[1] (b_n) where

$$b_n = a_{n+1} \cdot a_{n+1} - a_n - 1$$

To compute the limit of (b_n), use the multiplication and sum properties of limits:

$$b_n \to L \cdot L - L - 1$$

When we expand the right hand side of the original definition of b_n, we see that

$$b_n = \underbrace{1 + a_n}_{a_{n+1} \cdot a_{n+1}} - a_n - 1 = 0.$$

Therefore,

$$b_n \to 0$$

Since

$$b_n \to 0$$
$$b_n \to L \cdot L - 1 - L,$$

by **uniqueness of limits**,

$$0 = L \cdot L - 1 - L.$$

Solving the quadratic and taking the positive solution,

$$L = \frac{1 + \sqrt{5}}{2}. \qquad \square$$

10.3   The Sandwich Theorem

The Sandwich Theorem states that

If we have a sequence "sandwich-ed" between two sequences that converge to the same limit, then the "sandwich-ed" sequence converges to this limit as well.

Like the Monotone Convergence Property, the biggest application of the Sandwich Theorem is proving the Bolzano-Weierstrass Theorem (which, like Johnny Depp in any movie, is literally the star of the show).

The Sandwich Theorem, however, is important in its own right as it will allow us to prove

$$\sqrt[n]{n} \to 1.$$

This limit will be used in our proof of the *Change of Base-Point Theorem* in **Lecture 21**.

---

[1]The definition of b_n mirrors the bad proof's step of $y^2 = y - 1$ (or really, $y^2 - y + 1 = 0$). Even the bad proof has some good ideas to learn from!

Theorem (Sandwich Theorem). *If*

$$a_i \leq b_i \leq c_i$$

for all i and

$$a_n \ \to \ L$$
$$c_n \ \to \ L$$

then (b_n) converges and

$$b_n \to L.$$

Proof Summary:

1. Let $\epsilon > 0$.

2. There are N_1, N_2 such that all terms in (a_n) and (b_n) beyond the $N_1 - th$ term and $N_2 - th$ term, respectively, are within ϵ of L.

3. Let $N = \max\{N_1, N_2\}$.

4. Explicitly write out all the limit definitions as inequalities without absolute values, and combine.

Proof: Let $\epsilon > 0$. We want to show there is some N such that for all $i \geq N$,

$$|b_i - L| < \epsilon$$

When we expand the absolute value, this means that we have to prove the pair of inequalities:

$$L - \epsilon \ < \ b_i$$
$$b_i \ < \ L + \epsilon$$

By the definition of convergence for (a_n), there exists an N_1 such that for all $i \geq N_1$

$$|a_i - L| < \frac{\epsilon}{2}.$$

This inequality is equivalently

$$\left. \begin{array}{ccc} L - \epsilon & < & a_i \\ a_i & < & L + \epsilon \end{array} \right\} (\star)$$

Likewise, we know there exists an N_2 such that for all $i \geq N_2$,

$$\left. \begin{array}{ccc} L - \epsilon & < & c_i \\ c_i & < & L + \epsilon \end{array} \right\} (\star\star)$$

Since we want both $(\star)$ and $(\star\star)$ to hold, let

$$N = \max\{N_1, N_2\}$$

Then, for all $i \geq N$,

$$L - \epsilon \;\; < \;\; a_i \quad \text{by } (\star)$$
$$\leq \;\; b_i$$

Likewise,

$$b_i \;\; \leq \;\; c_i$$
$$< \;\; L + \epsilon \quad \text{by } (\star\star)$$

Thus, for all $i \geq N$,

$$L - \epsilon < b_i < L + \epsilon. \qquad \qquad \Box$$

To prove

$$\sqrt[n]{n} \to 1,$$

we need one quick result.

Lemma. *If $x \geq 1$, then*

$$\sqrt[n]{x} \geq 1.$$

Proof: Suppose $\sqrt[n]{x} < 1$. The product of positive numbers less than 1 is still less than 1 (recall the ordering axioms). Therefore,

$$\overbrace{\underbrace{\sqrt[n]{x}}_{<1} \cdot \underbrace{\sqrt[n]{x}}_{<1} \cdots \cdot \underbrace{\sqrt[n]{x}}_{<1}}^{n-times} < 1 \cdot 1 \cdot \ldots \cdot 1 = 1$$

But we have n copies of $\sqrt[n]{x}$ on the left hand side, so

$$x < 1$$

a contradiction. $\qquad\qquad \Box$

Now, we are ready.

Theorem.

$$\sqrt[n]{n} \to 1$$

Proof Summary:

- Define sequence $s_n = \sqrt[n]{n} - 1$.

- $(1 + s_n)^n$ bounds the *third term* of its binomial expansion.

- Isolate s_n and use the Sandwich Theorem to show the limit of s_n is 0.

- Conclude $(\sqrt[n]{n}) \to 1$ by the limit sum property.

Proof: First, to make everything look nicer, define

$$s_n = \sqrt[n]{n} - 1.$$

By the preceding theorem, since $n \geq 1$,

$$s_n \geq 0.$$

Now here's the trick: consider

$$(1 + s_n)^n.$$

By the awesomeness of Binomial Theorem,

$$(1 + s_n)^n = \underbrace{\binom{n}{0} 1^n (s_n)^0}_{\geq 0} + \underbrace{\binom{n}{1} 1^{n-1} (s_n)^1}_{\geq 0} + \ldots + \underbrace{\binom{n}{n} 1^0 (s_n)^n}_{\geq 0}.$$

All these terms are non-negative since s_n is non-negative (that's why we needed the previous theorem)! Therefore, $(1 + s_n)^n$ must be greater than (or equal to) each individual term in its sum. In particular, it bound the *third* term:

$$(1 + s_n)^n \geq \underbrace{\frac{n \cdot (n-1)}{2} (s_n)^2}_{\binom{n}{2} 1^{n-2} (s_n)^2}$$

Substituting s_n into the (LHS)

$$\underbrace{n}_{(1+s_n)^n} \geq \frac{n \cdot (n-1)}{2} (s_n)^2$$

and simplifying,

$$s_n \leq \sqrt{\frac{2}{n-1}}.$$

Again, since s_n is non-negative,

$$0 \leq s_n \leq \sqrt{\frac{2}{n-1}}.$$

Since the zero sequence converges to 0 and we can easily show

$$\sqrt{\frac{2}{n-1}} \to 0,$$

we can conclude

$$s_n \to 0$$

by the Sandwich Theorem. Therefore, by the limit sum theorem,[1]

$$s_n + 1 \to 1.$$

In other words,

$$\sqrt[n]{n} \to 1. \qquad \square$$

On this week's homework, you will take the Sandwich Theorem one step further to prove that if

$$\frac{1}{n^k} \leq a_n \leq n^k$$

for every n and some positive integer k, then

$$\lim_{n \to \infty} \sqrt[n]{a_n} = 1.$$

---

[1]Note, the 1 signifies a *constant sequence* $c_n = 1$ for all n.

10.4 Bolzano-Weierstrass Theorem

To explain the proof of *Bolzano − Weierstrass*, let's play a game: the game of *bisection*. If you've ever been to nerd-camp, you've probably seen this game before. Some cocky little trickster, say Puck, asserts

> *Puck: I betcha I can guess your birthday in 9 tries, as long as, after each try, you tell me if your birthday is later or earlier.*

And if you're a naive little Muggle, you might reply:

> *Muggle: You're on!*

Say Muggle's birthday is September 4th. Puck first thinks of all the dates of the year as an interval:

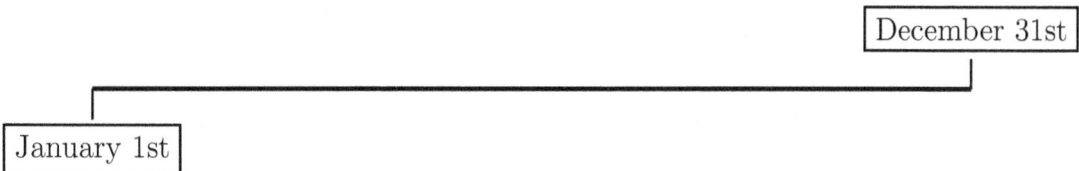

In order to eliminate as many choices as possible, Puck asks Muggle to compare her birthday to the middle of the year:

> *Puck: Is your birthday before or after July 1st?*
> *Muggle: After*

So Puck erases all the dates before July 1st.

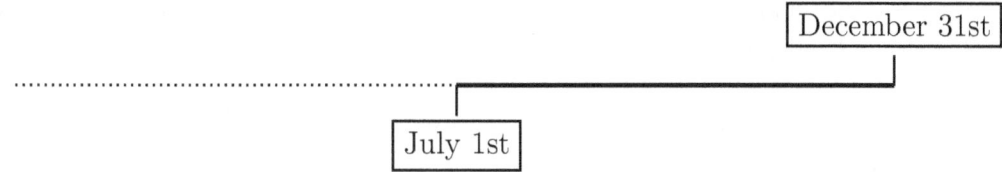

He then calculates the new middle of this interval, October 1st.

> *Puck: Is your birthday before or after October 1st?*
> *Muggle: Before*

Then he removes all dates after October 1st.

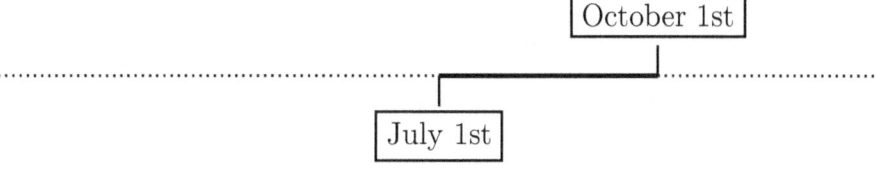

The game continues for a few more iterations.

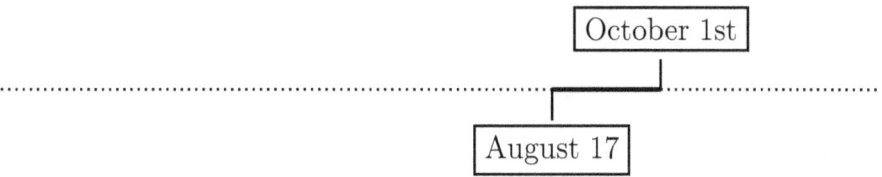

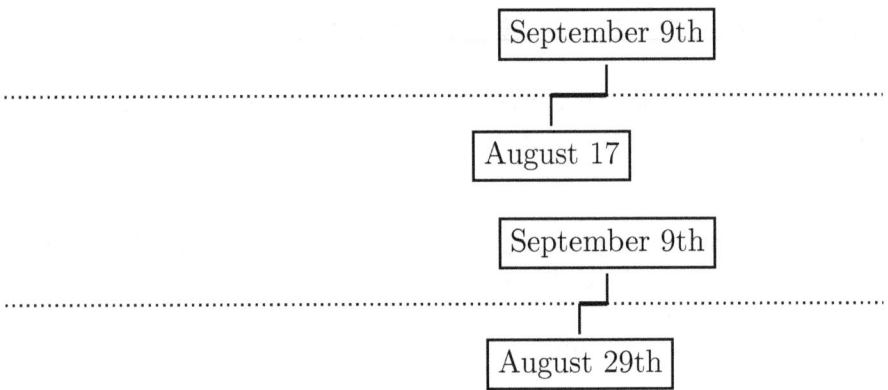

Finally, on the sixth guess,

> *Puck: Is your birthday September 4th?*
> *Muggle: Wow you're psychic!*

Pretty simple, huh? Bolzano-Weierstrass constructs the subsequence in the *same* way! However, there are a few twists:

- The game goes on forever.

- "Guesses" will be terms in the original sequence.

- Instead of choosing the interval containing the birthday, you choose the interval that contains infinitely many terms of the original sequence. This ensures the game *never ends*.

It turns out that by the awesomeness of Monotone Convergence Property and Sandwich Theorem, your sequence of guesses is convergent!

Theorem (Bolzano-Weierstrass). *Every bounded sequence has a convergent subsequence.*

Proof Summary:

- Define a sequence of intervals, each one half the size of the previous and each containing infinitely many points of (a_n)

- From the j-th interval, select a term to be a_{n_j}.

- Use the Monotone Convergence Theorem to prove that the sequence of left interval endpoints and the sequence of right interval endpoints both converge.

- Use the Sandwich Theorem on the endpoint sequences to show (a_{n_j}) converges.

Proof: Let (a_n) be bounded by K. So that means we are starting with the interval

$$I_1 = [-K, K]$$

To help us understand the process, we are going to use the following schematic:

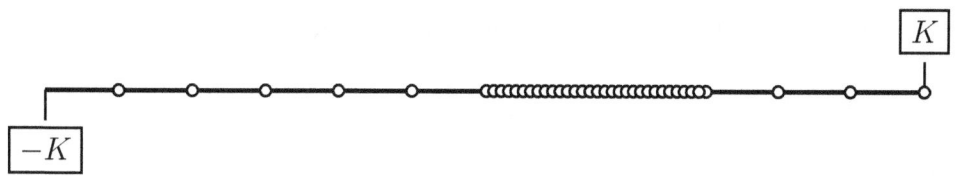

Here, each dot represents a different term of the sequence. Choose any a_{n_1} in this interval. In particular, we can choose

$$a_{n_1} = a_1.$$

Now consider the two subintervals

$$[-K, 0] \quad [0, K]$$

Since a sequence is infinite, at least[1] **one of these interval contains infinitely many terms.** Let I_2 be this interval.

In our schematic,

$$I_2 = [0, K]$$

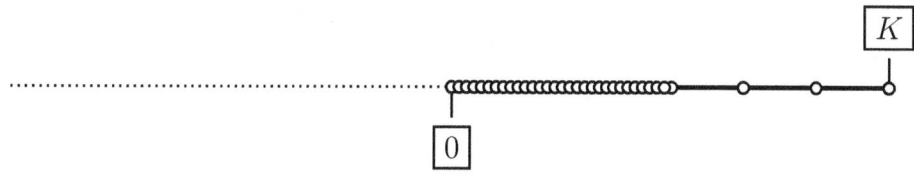

Now choose a point $a_{n_2} \in I_2$. Make sure that you choose an **index that is bigger than the previous**:

$$n_2 > n_1.$$

This is because subsequences must have *increasing* indices. And because our interval contains *infinitely* many sequence terms, we know such an n_2 *exists*.

Now we have

$$a_{n_1} \in I_1$$
$$a_{n_2} \in I_2$$

Continuing, split interval I_2 in half and let I_3 be the interval that contains infinitely many terms of (a_n). Choose a_{n_3} to be a point in this interval, such that

$$n_3 > n_2$$

In our schematic, we choose

$$I_3 = \left[0, \frac{K}{2}\right]$$

[1]You may have a choice between two intervals containing infinitely many terms. Just pick one, but you should see from this that there may be many *different* convergent subsequences, each converging to a different limit. Mull over why this doesn't contradict the fact that any subsequence of a *convergent* sequence converges to the same limit.

Now that you understand the intuition, we give the inductive definition:

The intervals I_n are defined[1] as

$$I_1 = [-K, K]$$

$$I_n = \begin{cases} \left[a, \frac{a+c}{2}\right] & : \text{ if } I_{n-1} = [a, c] \text{ and the interval } \left[a, \frac{a+c}{2}\right] \text{ contains infinitely many sequence terms} \\ \left[\frac{a+c}{2}, c\right] & : \text{ else} \end{cases}$$

The inductive definition of the subsequence is just

$$a_{n_1} = a_1$$
$$a_{n_j} = \text{ some point of } (a_n) \text{ in } I_j \text{ such that } n_j > n_{j-1}$$

To reiterate, we know a_{n_j} exists since I_j contains infinitely many terms of (a_n) by construction.

I claim (a_{n_j}) converges. And to prove this, we will use the Sandwich Theorem. Namely, we will use the fact that our sequence is sandwiched between the sequences formed from the **endpoints of the intervals**.

Let
$$c_j = \text{ Left endpoint of interval } I_j$$
$$d_j = \text{ Right endpoint of interval } I_j$$

We check the preconditions of the Sandwich Theorem:

- $c_j \leq d_j$
  They are the left and right endpoints of interval I_j, duh.

- (c_j) **and** (d_j) **both converge.**
  Inductively, you can show that (c_j) is a *monotonic increasing sequence*: every time you cut the interval in two, your next endpoint either

  - Stays the same

  - Is the midpoint of the previous interval.

  Likewise, we can prove (d_j) is *monotonic decreasing*. Since they are both bounded by K, we know by the Monotone Convergence Theorem, $(c_j), (d_j)$ both converge and their limits exist.

- (c_j) **and** (d_j) **converge to the same limit.**
  Let
  $$c_j \rightarrow L_1$$
  $$d_j \rightarrow L_2$$

  To prove $L_1 = L_2$, it suffices to show

  $$(d_j - c_j) \rightarrow 0.$$

---

[1]Don't be scared of the whole $\frac{a+c}{2}$ business: this is just the *midpoint* of a and c.

This is because, by our limit sum properties,

$$\underbrace{(d_i - c_i}_{\to 0} + \underbrace{c_i}_{\to L_1}) \to L_1$$

Moreover,

$$(\underbrace{d_i}_{\to L_2} - \underbrace{c_i + c_i}_{=0}) \to L_2$$

By uniqueness,

$$L_2 = L_1.$$

Showing

$$(d_i - c_i) \to 0$$

is easy: it is just the *interval length*. From our construction, each interval is *half* the size of the previous. Inductively,

$$d_i - c_i = \frac{2K}{2^{i-1}}$$

which converges to 0.

In conclusion, by the Sandwich Theorem on

$$c_j \le a_{n_j} \le d_j,$$

(a_{n_j}) converges. Awesome. $\qquad\qquad\qquad\qquad\qquad\qquad\qquad\qquad\qquad\qquad\quad\square$

By the way, there are two other infamous proofs of Bolzano-Weierstrass involving

- *limsups*

- *dominated terms*

I leave the first item for **Math 171**. The second is a *very cute* forty second proof.

10.5 An Alternate Proof of Bolzano-Weierstrass

In a given sequence (a_n), we say that a term a_k is *dominant* if it is strictly greater than all the terms that appear later in the sequence:

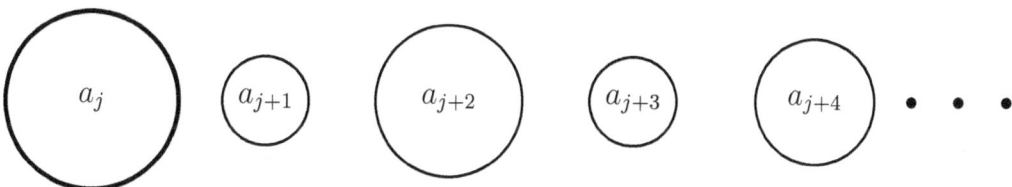

Formally,

Definition. *Let (a_n) be a sequence. Then we say a term a_k is* **dominant** *if for all $i > k$,*

$$a_k > a_i.$$

Otherwise, there exists some $i > k$ such that

$$a_i \geq a_k,$$

and we say a_k is **conquerable** *and that a_i* **dominates** *a_k.*

Using this jargon, we can give a very simple proof of the Bolzano-Weierstrass theorem:

Theorem (Bolzano-Weierstrass). *Every bounded sequence has a convergent subsequence.*

Proof Summary:

- There are two possible cases:

    - THERE ARE INFINITELY MANY DOMINANT TERMS.
        * The subsequence of *dominant* terms is *monotonically decreasing.*
        * Apply the Monotone Convergence Property.
    - THERE ARE FINITELY MANY DOMINANT TERMS.
        * After the last dominant term, every term is *conquerable.*
        * Construct a subsequence of conquerable terms such that each term is dominated by the next term. This is a *monotonically increasing subsequence.*
        * Apply the Monotone Convergence Property.

Proof: We have one of two cases. Either:

THERE ARE INFINITELY MANY DOMINANT TERMS

OR

THERE ARE FINITELY MANY DOMINATED TERMS

- THERE ARE INFINITELY MANY DOMINANT TERMS.
  Define (a_{n_i}) where

$$a_{n_i} = \text{the } i\text{-th dominant term.}$$

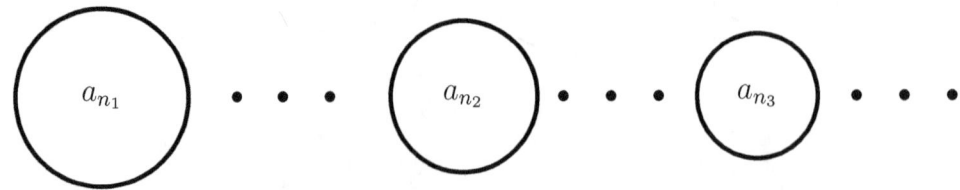

By definition, this is a *monotonically decreasing sequence*. Since it's also bounded, it follows by the Monotone Convergence Property that (a_{n_i}) converges.

- THERE ARE FINITELY MANY DOMINANT TERMS.
  Since there are *finitely* many dominant terms, there is a *last* dominant term, say a_N. Then all the terms after a_N are conquerable. Formally, for all $i > N$, there exists some $j > i$ such that

$$a_i \leq a_j.$$

Construct a subsequence (a_{n_i}) where each next term is *dominated* by the next:[1]

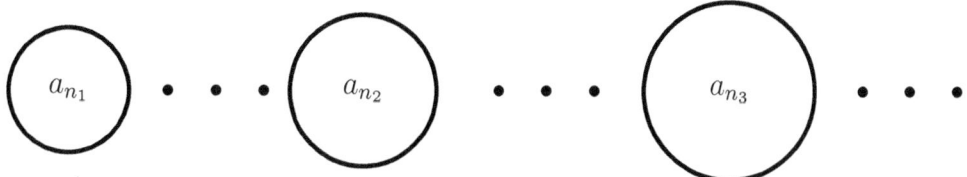

In particular, choose

$$a_{n_1} = a_{N+1}.$$

Then for each $i \geq 1$, supposing a_{n_i} has been selected, we may select

$$a_{n_{i+1}} = a_j \text{ for some } a_j \text{ that dominates } a_{n_i}.$$

By construction, (a_{n_i}) is a *monotonically increasing sequence*. Since it's also bounded, it follows by the Monotone Convergence Property that (a_{n_i}) converges.

In both cases, we constructed a convergent subsequence. $\square$

---

[1]Think of a fish being eaten by a bigger fish and that fish being eaten by an even bigger fish, and so on ad infinitum.

Lecture 11

Fishing for Complements

They go together like oil and water. Nothing in common 'cept the zero vector.

- Mathematical idiom

Goals: We fulfill our promise and prove that the dimensions of the row space and the column space are equal. The key to this proof lies in improving the original proof of the Rank-Nullity Theorem. This will require introducing orthogonal complements and proving many of their properties. Afterwards, we use these properties to define projection maps.

11.1   The Story so Far...

Last week, we left you with a cliffhanger: the phenomenal fact that the dimensions of the row space and column space are equal:

$$\dim C(A) = \dim R(A)$$

Unlike *Lost*, we are not going to leave you stranded. To prove this fact, we need to take another look at the Rank-Nullity Theorem.

In truth, our proof of the Rank-Nullity theorem was *too fast*. If we took the time to *examine our steps more carefully*, we could have improved this theorem **tenfold.**

Consider the first step of the proof. Starting with a basis for the null space, we extended this basis to a full basis for the domain $\mathbb{R}^n$:

$$\underbrace{\vec{N}_1, \vec{N}_2, \ldots, \vec{N}_{n-q}}_{null\ space}, \underbrace{\vec{x}_1, \vec{x}_2, \ldots, \vec{x}_q}_{extension\ vectors}$$

At that point, we had **many choices** of extension vectors. In fact, we could have chosen them to have additional *structure*:

Math Mantra: If you have multiple choices for an object, choose one that has
additional structure.   Then EXPLOIT this extra structure in your proof.

But what structure should we impose on

$$\vec{x}_1, \vec{x}_2, \ldots, \vec{x}_q?$$

To determine this structure, we first need to ask ourselves,

How does the row space relate to the null space?

Write out a matrix multiplication for a vector in the null space and stare at it:

$$\begin{bmatrix} 1 & 0 & 1 \\ 0 & 1 & 1 \\ 0 & 0 & 0 \end{bmatrix} \begin{bmatrix} 1 \\ 1 \\ -1 \end{bmatrix} = \begin{bmatrix} 0 \\ 0 \\ 0 \end{bmatrix}$$

Notice that the null space vector has dot product 0 with each of the rows!

$$\begin{bmatrix} 1 \\ 0 \\ 1 \end{bmatrix} \cdot \begin{bmatrix} 1 \\ 1 \\ -1 \end{bmatrix} = 0 \qquad \begin{bmatrix} 0 \\ 1 \\ 1 \end{bmatrix} \cdot \begin{bmatrix} 1 \\ 1 \\ -1 \end{bmatrix} = 0 \qquad \begin{bmatrix} 0 \\ 0 \\ 0 \end{bmatrix} \cdot \begin{bmatrix} 1 \\ 1 \\ -1 \end{bmatrix} = 0$$

In fact, this holds for any vector in the null space *by definition*. This property is so important that we give it a name:

Definition. *We say $\vec{x}$ is **orthogonal** to $\vec{y}$ if*

$$\vec{x} \cdot \vec{y} = 0.$$

*For a subspace $V \subseteq \mathbb{R}^n$, we define the **orthogonal complement**[1] $V^\perp$ to be the set of all vectors that are orthogonal to every vector in V:*

$$V^\perp = \left\{ \vec{x} \in \mathbb{R}^n \,\middle|\, \vec{x} \cdot \vec{y} = 0 \text{ for all } \vec{y} \in V \right\}.$$

First, I claim

> CLAIM: We can choose the extension vectors
>
> $$\vec{x}_1, \vec{x}_2, \ldots, \vec{x}_q$$
>
> such that they are a basis for $\left(N(A) \right)^\perp$ and
>
> $$\underbrace{\vec{N}_1, \vec{N}_2, \ldots, \vec{N}_{n-q}}_{N(A)}, \underbrace{\vec{x}_1, \vec{x}_2, \ldots, \vec{x}_q}_{(N(A))^\perp}$$
>
> is still a basis for $\mathbb{R}^n$.

---

[1]Careful! Orthogonal complement is **not** the same as set complement!

I also claim that:

CLAIM: The orthogonal complement of the null space is the row space

$$\left(N(A)\right)^{\perp} = R(A)$$

So by our claims, we can in fact choose

$$\vec{x}_1, \vec{x}_2, \ldots, \vec{x}_q$$

to be a *basis for the row space*. Since we've already proven

$$A\vec{x}_1, A\vec{x}_2, \ldots, A\vec{x}_q$$

is a basis for the column space, we instantly have

$$\dim C(A) = \dim R(A)$$

Awesome!

Now that we have the game plan, let's verify these claims.

11.2 Proving the First Claim

Consider our first claim,

CLAIM: We can choose the extension vectors

$$\vec{x}_1, \vec{x}_2, \ldots, \vec{x}_q$$

such that they are a basis for $\left(N(A)\right)^{\perp}$ and

$$\underbrace{\vec{N}_1, \vec{N}_2, \ldots, \vec{N}_{n-q}}_{N(A)}, \underbrace{\vec{x}_1, \vec{x}_2, \ldots, \vec{x}_q}_{(N(A))^{\perp}}$$

is still a basis for $\mathbb{R}^n$

Instead of proving this claim directly, it is better to abstract (you'll see why)!

REVISED CLAIM: Given a subspace $V \subseteq \mathbb{R}^n$, let

$$\vec{v}_1, \vec{v}_2, \ldots \vec{v}_q$$

$$\vec{w}_1, \vec{w}_2, \ldots \vec{w}_r$$

be bases for $V, V^{\perp}$ respectively. Then,

$$\vec{v}_1, \vec{v}_2, \ldots, \vec{v}_q, \vec{w}_1, \vec{w}_2, \ldots \vec{w}_r$$

is a basis for $\mathbb{R}^n$.

This is a better claim, but we still need to be careful! Our claim could be **completely nonsensical**: we forgot to check that $V^\perp$ is the right *animal*. Remember, for our discussion of bases to make sense, $V^\perp$ needs to be a subspace! Let's verify this:

Lemma. *Given subspace $V \subseteq \mathbb{R}^n$, $V^\perp$ is also a subspace.*

Proof:

- *Existence of Zero*
  For any $\vec{x} \in V$,
  $$\vec{x} \cdot \vec{0} = 0.$$
  Hence,
  $$\vec{0} \in V^\perp.$$

- *Closure under Addition*
  Let $\vec{a}, \vec{b} \in V^\perp$. Then for any $\vec{x} \in V$
  $$\vec{x} \cdot (\vec{a} + \vec{b}) = \underbrace{\vec{x} \cdot \vec{a}}_{0} + \underbrace{\vec{x} \cdot \vec{b}}_{0} = 0.$$
  Thus,
  $$(\vec{a} + \vec{b}) \in V^\perp.$$

- *Closure under Scaling*
  Let $\vec{a} \in V^\perp$ and $k \in \mathbb{R}$. Then for any $\vec{x} \in V$,
  $$\vec{x} \cdot (k\vec{a}) = k(\underbrace{\vec{x} \cdot \vec{a}}_{0}) = 0$$
  so
  $$k\vec{a} \in V^\perp. \qquad \square$$

To prove our revised claim, we need to prove a few fundamental facts about subspaces.

First, we need to show that V and $V^\perp$ have only the zero vector in common:

Lemma. *For any subspace V,*
$$V \cap V^\perp = \{\vec{0}\}.$$

Proof:

- $\supseteq$
  Since V and $V^\perp$ are vector spaces,
  $$\vec{0} \in V \cap V^\perp.$$

- $\subseteq$

Now let $\vec{x} \in V \cap V^{\perp}$ i.e

$$\vec{x} \in V$$
$$\vec{x} \in V^{\perp}.$$

Since $\vec{x} \in V$,

$$\vec{x} \cdot \vec{y} = 0$$

for any $\vec{y} \in V^{\perp}$. But $\vec{x} \in V^{\perp}$, so **choose** in particular

$$\vec{y} = \vec{x}$$

Then,

$$\vec{x} \cdot \underbrace{\vec{x}}_{\vec{y}} = 0$$

i.e

$$\|\vec{x}\|^2 = 0$$

A vector has norm zero if and only it is the zero vector; thus,

$$\vec{x} = \vec{0}. \qquad \square$$

The following theorem states that the whole space, and only the whole space, has just the zero vector in its orthogonal complement.

Lemma. *For a vector space $V \subseteq \mathbb{R}^n$,*

$$V^{\perp} = \{\vec{0}\} \iff V = \mathbb{R}^n$$

Proof Summary:

- $\Leftarrow$

    - $\vec{x} \cdot \vec{e}_i = 0$ for all i.
    - $\vec{x} \cdot \vec{e}_i = x_i$. Thus, $\vec{x} = \vec{0}$.

- $\Rightarrow$

    - Suppose the dimension of V is less than n.
    - Taking the dot product of an arbitrary $\vec{x}$ with each of $V's$ basis vectors yields a system with fewer equations than unknowns.
    - This system has a non-trivial solution by Under-determined Systems Lemma.
    - This non-trivial solution is in $V^{\perp}$, contradiction.

Proof:

- $\Leftarrow$

  We are given $V = \mathbb{R}^n$. Let $\vec{x} \in V^{\perp}$. Then for any $\vec{v} \in V$,

  $$\vec{x} \cdot \vec{v} = 0$$

  Choose an arbitrary standard basis $\vec{e}_i$, and let $\vec{v} = \vec{e}_i$:

  $$\vec{x} \cdot \underbrace{\vec{e}_i}_{\vec{v}} = 0$$

  Dotting with $\vec{e}_i$ picks out the i-th component:

  $$\begin{bmatrix} x_1 \\ \vdots \\ x_{i-1} \\ x_i \\ x_{i+1} \\ \vdots \\ 0 \end{bmatrix} \cdot \begin{bmatrix} 0 \\ \vdots \\ 0 \\ 1 \\ 0 \\ \vdots \\ 0 \end{bmatrix} = x_i.$$

  This tells us that the i-th component is 0:

  $$x_i = 0$$

  Of course, since this is true for any i,
  $$\vec{x} = \vec{0}$$

  and hence
  $$V^{\perp} \subseteq \{\vec{0}\}.$$

  Of course, $\{\vec{0}\} \subseteq V^{\perp}$, so we conclude
  $$V^{\perp} = \{\vec{0}\}.$$

- $\Rightarrow$

  Now we are given $V^{\perp} = \{\vec{0}\}$. We need to show $V = \mathbb{R}^n$.

  Suppose not. Then V has dimension $k < n$ and we can select a basis

  $$\vec{v}_1, \vec{v}_2, \ldots \vec{v}_k.$$

  Consider the dot product of each v_i with an arbitrary vector $\vec{x} \in \mathbb{R}^n$

  $$\begin{aligned} \vec{v}_1 \cdot \vec{x} &= 0 \\ \vec{v}_2 \cdot \vec{x} &= 0 \\ &\vdots \\ \vec{v}_k \cdot \vec{x} &= 0 \end{aligned}$$

Expanding each line, this gives us a system of k equations with n unknowns. By the Underdetermined Systems Lemma, this system has a non-trivial solution $\vec{x}_0 \neq \vec{0}$. But we can show this non-trivial solution $\vec{x}_0$ is in $V^\perp$. For any $\vec{v} \in V$, we can write

$$\vec{v} = a_1 \vec{v}_1 + a_2 \vec{v}_2 + \ldots + a_k \vec{v}_k.$$

Then,

$$
\begin{aligned}
\vec{x}_0 \cdot \vec{v} &= \vec{x}_0 \cdot \underbrace{\left(a_1 \vec{v}_1 + a_2 \vec{v}_2 + \ldots + a_k \vec{v}_k\right)}_{=\vec{v}} \\
&= a_1 \underbrace{\left(\vec{x}_0 \cdot \vec{v}_1\right)}_{=0} + a_2 \underbrace{\left(\vec{x}_0 \cdot \vec{v}_2\right)}_{=0} + \ldots + a_k \underbrace{\left(\vec{x}_0 \cdot \vec{v}_k\right)}_{=0} \\
&= 0
\end{aligned}
$$

Therefore $\vec{x}_0 \in V^\perp$, but $\vec{x}_0 \neq \vec{0}$; this contradicts the fact that $V^\perp$ *only* contains the zero vector! Thus,

$$V = \mathbb{R}^n. \qquad \square$$

For the next property, define

Definition. *The **sum** of subspaces $A, B \subseteq \mathbb{R}^n$ is defined as*

$$A + B = \{\vec{x} + \vec{y} \mid \vec{x} \in A, \vec{y} \in B\}$$

It is a very easy exercise to prove

$$A + B = \operatorname{span}\{\vec{\alpha}_1, \vec{\alpha}_2, \ldots, \vec{\alpha}_q, \vec{\beta}_1, \vec{\beta}_2, \ldots, \vec{\beta}_r\}$$

where

$$\vec{\alpha}_1, \vec{\alpha}_2, \ldots, \vec{\alpha}_q$$
$$\vec{\beta}_1, \vec{\beta}_2, \ldots, \vec{\beta}_s.$$

are bases for A, B respectively. In particular,

$$V + V^\perp = \operatorname{span}\{\vec{v}_1, \vec{v}_2, \ldots, \vec{v}_q, \vec{w}_1, \vec{w}_2, \ldots, \vec{w}_r\}$$

where the v's are a basis for V and the w's are a basis for $V^\perp$. Now we can prove part of our revised claim, namely that

$$\operatorname{span}\{\vec{v}_1, \vec{v}_2, \ldots, \vec{v}_q, \vec{w}_1, \vec{w}_2, \ldots, \vec{w}_r\} = \mathbb{R}^n,$$

by showing:

Lemma. *For any subspace $V \subseteq \mathbb{R}^n$,*

$$V + V^\perp = \mathbb{R}^n$$

Proof Summary:

- It suffices to prove that the complement of $V + V^\perp$ is the zero vector.

- Any vector $\vec{x} \in \left(V + V^\perp\right)^\perp$ is orthogonal to every vector in $V + V^\perp$.

- $V \subseteq V + V^\perp$ so $\vec{x}$ is orthogonal to every vector in V, hence $\vec{x} \in V^\perp$.

- Thus, $\vec{x} \in V + V^\perp$.

- $\vec{x}$ is in a subspace and its complement, so $\vec{x} = \vec{0}$.

Proof: By the previous lemma, it suffices to show that

$$\left(V + V^\perp\right)^\perp = \{\vec{0}\}$$

Of course, $\{\vec{0}\} \subseteq \left(V + V^\perp\right)^\perp$, so we need only show $\left(V + V^\perp\right)^\perp \subseteq \{\vec{0}\}$.

Let $\vec{x} \in \left(V + V^\perp\right)^\perp$. Then for every $\vec{y} \in V + V^\perp$

$$\vec{x} \cdot \vec{y} = 0.$$

In particular,

$$V \subseteq V + V^\perp$$

so for every $\vec{w} \in V$,

$$\vec{x} \cdot \vec{w} = 0.$$

This means $\vec{x} \in V^\perp$. Of course this implies

$$\vec{x} \in V + V^\perp$$

since

$$\vec{x} = \underbrace{\vec{0}}_{\in V} + \underbrace{\vec{x}}_{\in V^\perp}.$$

Now,

$$\begin{aligned} \vec{x} &\in \left(V + V^\perp\right)^\perp \\ \vec{x} &\in V + V^\perp \end{aligned}$$

But we proved that only the zero vector is in both a subspace and its complement, so

$$\vec{x} = \vec{0}. \qquad \square$$

Theorem (First Claim). *Given subspace $V \subseteq \mathbb{R}^n$, let*

$$\vec{v}_1, \vec{v}_2, \ldots \vec{v}_q$$
$$\vec{w}_1, \vec{w}_2, \ldots \vec{w}_r$$

be bases for $V, V^\perp$ respectively. Then,

$$\vec{v}_1, \vec{v}_2, \ldots, \vec{v}_q, \vec{w}_1, \vec{w}_2, \ldots \vec{w}_r$$

is a basis for $\mathbb{R}^n$.

Proof Summary:

- *Spanning*: Already proved $V + V^\perp = \mathbb{R}^n$.

- *Linear Independence:*

  - Suppose not. Then we have a non-trivial combination of $\vec{0}$.
  - Move all V terms on one side and $V^\perp$ to the other.
  - (LHS) is in V and (RHS) is in $V^\perp$. So both sides must be $\vec{0}$ since it is the only vector in V and $V^\perp$.
  - Use basis definition to conclude original combination was trivial
  - Contradiction.

Proof:

- *Spanning*: We already proved
$$V + V^\perp = \mathbb{R}^n.$$

- *Linear Independence*: Suppose

$$\underbrace{\vec{v}_1, \vec{v}_2, \ldots \vec{v}_q}_{V}, \underbrace{\vec{w}_1, \vec{w}_2, \ldots \vec{w}_r}_{V^\perp}$$

is not linearly independent. Then, there is a non-trivial combination that yields $\vec{0}$:

$$c_1 \vec{v}_1 + c_2 \vec{v}_2 + \ldots + c_q \vec{v}_q + c_{q+1} \vec{w}_1 + c_{q+2} \vec{w}_2 + \ldots + c_{q+r} \vec{w}_r = \vec{0}$$

Isolate all the V vectors on one side:

$$\underbrace{c_1 \vec{v}_1 + c_2 \vec{v}_2 + \ldots + c_q \vec{v}_q}_{V} = \underbrace{-c_{q+1} \vec{w}_1 - c_{q+2} \vec{w}_2 - \ldots - c_{q+r} \vec{w}_r}_{V^\perp}$$

This gives us a vector in V that is equal to a vector in $V^\perp$. But we proved there is *only one* vector that is **both** in V and $V^\perp$: the zero vector. Thus, we have

$$
\begin{aligned}
c_1 \vec{v}_1 + c_2 \vec{v}_2 + \ldots + c_q \vec{v}_q &= \vec{0} \\
-c_{q+1} \vec{w}_1 - c_{q+2} \vec{w}_2 - \ldots - c_{q+r} \vec{w}_r &= \vec{0}
\end{aligned}
$$

But the v's are a basis for V and the w's are a basis for $V^\perp$. Therefore,

$$
\begin{aligned}
c_1 = c_2 = \ldots = c_q &= 0 \\
c_{q+1} = c_{q+2} = \ldots = c_{q+r} &= 0
\end{aligned}
$$

which contradicts that the combination is non-trivial.

Thus we have a basis for $\mathbb{R}^n$. $\qquad\square$

Note that this immediately implies

Theorem. *Given a subspace* $V \subseteq \mathbb{R}^n$,

$$\dim V + \dim V^\perp = n.$$

Recall the second claim:

11.3   Proving the Second Claim

Now let's prove the second claim,

CLAIM: The orthogonal complement of the null space is the row space

$$\left(N(A)\right)^\perp = R(A)$$

Unfortunately, if we tried to directly prove this directly, we would get stuck. We can easily prove that any vector in the row space is orthogonal to every vector in the null space. But,

*How do we know that the row space is **all** vectors of* $\left(N(A)\right)^\perp$?

Instead, we try a different approach. Consider the easier lemma:

Lemma. *For any* $m \times n$ *matrix* A,
$$\left(R(A)\right)^\perp = N(A)$$

Proof: Let $\vec{A}_1, \vec{A}_2, \ldots, \vec{A}_m$ be the rows of A.

- $\subseteq$
  Let $\vec{x} \in \left(R(A)\right)^\perp$. Then
  $$\vec{x} \cdot \vec{y} = 0$$

  for every $\vec{y} \in R(A)$. In particular, since the transposed rows[1] of A are in $R(A)$:

  $$\vec{x} \cdot \vec{A}_i^T = 0$$

  for $i = 1, 2, \ldots m$.

---

[1] On the first day of Math 51H, Professor Simon states that, instead of arrows, you should denote vectors by *underline*. Here, you can see why.

Representing this as a matrix multiplication,

$$
\begin{bmatrix}
\text{------} & \vec{A}_1 & \text{------} \\
\text{------} & \vec{A}_2 & \text{------} \\
 & \vdots & \\
\text{------} & \vec{A}_m & \text{------}
\end{bmatrix}
\begin{bmatrix}
x_1 \\ x_2 \\ \vdots \\ x_n
\end{bmatrix}
=
\begin{bmatrix}
0 \\ 0 \\ \vdots \\ 0
\end{bmatrix}
$$

we can see $\vec{x} \in N(A)$.

- $\supseteq$

  Let $\vec{x} \in N(A)$. We want to show

$$
\vec{x} \cdot \vec{r} = 0
$$

for any $\vec{r} \in R(A)$. Any vector $\vec{r} \in R(A)$ is a span of the transposed row vectors:

$$
\vec{r} = r_1 \vec{A}_1^T + r_2 \vec{A}_2^T + \ldots + r_n \vec{A}_n^T
$$

Again, we know

$$
\vec{x} \cdot \vec{A}_i^T = 0
$$

for $i = 1, 2, \ldots m$. Thus,

$$
\begin{aligned}
\vec{x} \cdot \vec{r} &= \vec{x} \cdot \underbrace{r_1 \vec{A}_1^T + r_2 \vec{A}_2^T + \ldots + r_n \vec{A}_n^T}_{\vec{r}} \\
&= \underbrace{r_1(\vec{x} \cdot \vec{A}_1^T)}_{=0} + \underbrace{r_2(\vec{x} \cdot \vec{A}_2^T)}_{=0} + \ldots + \underbrace{r_n(\vec{x} \cdot \vec{A}_n^T)}_{=0} \\
&= 0
\end{aligned}
$$

which implies $\vec{x} \in \left(R(A) \right)^\perp$. $\square$

Since the two sides are identical, we can complement both sides:

$$
\left((R(A))^\perp \right)^\perp = \left(N(A) \right)^\perp.
$$

Now, if we could "cancel out" the double complement, we'd be done:

$$
R(A) = \left(N(A) \right)^\perp
$$

Thus, we just need one more orthogonal complement property: the complement of the complement is the original subspace.

Lemma.

$$
V = \left(V^\perp \right)^\perp
$$

Proof Summary:

- By homework, suffices to prove V is contained in $\left(V^{\perp}\right)^{\perp}$ and they have the same dimensions:

- $\subseteq$

  – Directly from definition.

- $\dim V = \dim \left(V^{\perp}\right)^{\perp}$

  – Substitute V for S in $\dim S + \dim S^{\perp} = n$.
  – Substitute $V^{\perp}$ for S in $\dim S + \dim S^{\perp} = n$.
  – Combine the equations.

Proof:

- $\subseteq$
  Let $\vec{v} \in V$. To prove

$$\vec{v} \in \left(V^{\perp}\right)^{\perp}$$

  we have to show that, for any $\vec{x} \in V^{\perp}$,

$$\vec{x} \cdot \vec{v} = 0.$$

  But because $\vec{v} \in V$, this follows immediately from the definition of the complement space V^{perp}

Unfortunately, the $\supseteq$ direction is not as easy. If you tried to prove it directly, you would get stuck. Instead,

```
    Math Mantra:  If you cannot prove a theorem using the typical methods, then
      exploit the structure of the objects so that a different method will work.
```

Recall the homework exercise

Homework. *For subspaces A, B if*
$$A \subseteq B$$
and
$$\dim A = \dim B$$
then $A = B$.

Intuitively, this says that if one subspace is within another and both these subspaces have the same size, then they **must be equal**. Instead of proving "$\supseteq$" separately, we can **exploit the additional subspace structure of our sets**. Therefore, it suffices to prove that the dimensions of V and $\left(V^{\perp}\right)^{\perp}$ are equal:

- $\dim V = \dim \left(V^{\perp}\right)^{\perp}$

  From the first claim, we concluded[1]

  $$\dim S + \dim S^{\perp} = n$$

  for any subspace S. Plugging in V for S gives:

  $$\dim V + \dim V^{\perp} = n.$$

  Moreover, plugging in $V^{\perp}$ for S yields:

  $$\dim V^{\perp} + \dim \left(V^{\perp}\right)^{\perp} = n.$$

  Equating, we get

  $$\dim V = \dim \left(V^{\perp}\right)^{\perp} \qquad \square$$

The second claim is now easy to prove:

Theorem (Second Claim).

$$\left(N(A)\right)^{\perp} = R(A).$$

Proof: We already proved

$$\left(R(A)\right)^{\perp} = N(A)$$

Complement both sides

$$\left(\left(R(A)\right)^{\perp}\right)^{\perp} = \left(N(A)\right)^{\perp}$$

and cancel out the double complement:

$$R(A) = \left(N(A)\right)^{\perp}. \qquad \square$$

11.4   A Happy Ending

As mentioned in the introduction, we could easily use our proven claims to *rewrite* the proof of the Rank-Nullity Theorem so that it encodes

$$\dim C(A) = \dim R(A).$$

As a result of our hard work, however, there is no need. Instead, we can keep our original proof of the Rank-Nullity Theorem and *apply* our claims:

Theorem. *For any $m \times n$ matrix A,*

$$\dim C(A) = \dim R(A)$$

---

[1]Notice, we use a new dummy variable S since V is already in use.

Proof: We proved that for any subspace $V \subseteq \mathbb{R}^n$,

$$\dim V + \dim V^\perp = n,$$

so in particular,

$$\dim \underbrace{N(A)}_{V} + \dim \underbrace{\left(N(A)\right)^\perp}_{V^\perp} = n.$$

Using the fact

$$R(A) = \left(N(A)\right)^\perp,$$

our equality becomes

$$\dim N(A) + \dim R(A) = n.$$

Since the *original* Rank-Nullity Theorem gives us

$$\dim N(A) + \dim C(A) = n$$

we have

$$\dim N(A) + \dim R(A) = \dim N(A) + \dim C(A).$$

Hence,

$$\dim C(A) = \dim R(A). \qquad \square$$

11.5   Something Extra for our Troubles: Orthogonal Projection Map

We didn't need to talk generally about V and $V^\perp$ for any subspace V. Instead, we could have just stuck with $R(A)$ and $C(A)$. But,

> Math Mantra: By taking the extra time to abstract, we are rewarded.

Particularly, we now have the orthogonal complement properties:

ORTHOGONAL COMPLEMENT PROPERTIES

- A subspace and its complement only share the zero vector:
$$V \cap V^\perp = \vec{0}$$

- The sum of a subspace and its complement is the entire space:
$$V + V^\perp = \mathbb{R}^n$$

- The dimension of a subspace and its complement sum to the dimension of the full space:
$$\dim V + \dim V^\perp = n$$

- The complement of the complement is the original subspace:
$$\left(V^\perp\right)^\perp = V$$

Now that we have these neat properties, let's try to use them!

There were several times in high school when you needed to break a vector into orthogonal components. For example, in Algebra II you took a vector in the Cartesian Plane and divided it into a component along the x-axis and a component along the y-axis.

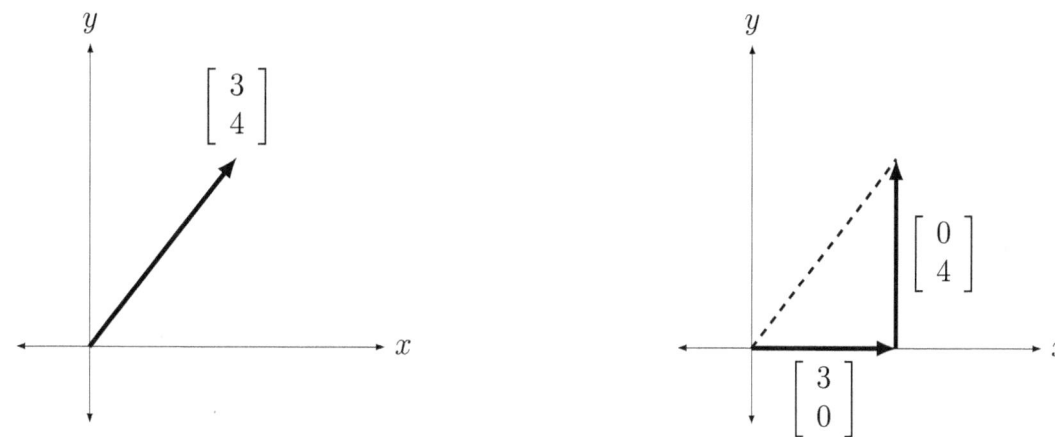

Another example is in Physics, when you studied force diagrams. For a sliding block, you broke the force of gravity into two orthogonal components. One component was normal to the surface, and the other component was parallel to the surface:

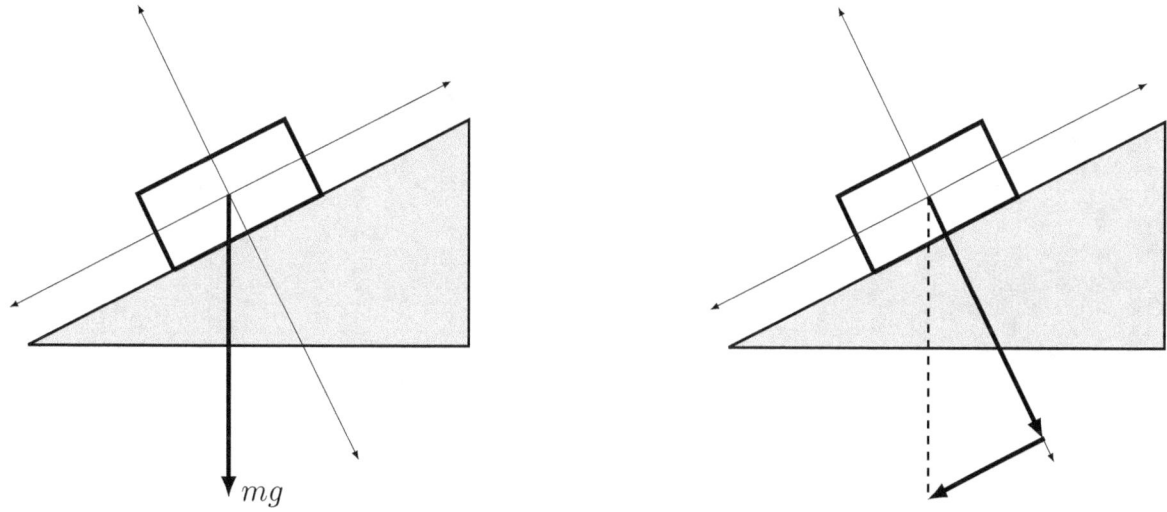

Today, we generalize these results! Namely, we proved for any subspace $V \subseteq \mathbb{R}^n$, we can break any vector $\vec{v} \in \mathbb{R}^n$ into a sum of two component vectors, one in the subspace and the other in the orthogonal complement:

$$\vec{x} = \underbrace{\vec{v}}_{\in V} + \underbrace{\vec{w}}_{\in V^\perp}$$

In fact, this decomposition is *unique*:

Theorem. *For any subspace $V \subseteq \mathbb{R}^n$ and vector $\vec{v} \in V$, if*

$$\vec{x} = \vec{v}_1 + \vec{w}_1$$

for some $\vec{v}_1 \in V$ and $\vec{w}_1 \in V^\perp$ and

$$\vec{x} = \vec{v}_2 + \vec{w}_2,$$

for some $\vec{v}_2 \in V$ and $\vec{w}_2 \in V^\perp$, then

$$\begin{aligned} \vec{v}_1 &= \vec{v}_2 \\ \vec{w}_1 &= \vec{w}_2 \end{aligned}$$

Proof: Again, we will use the trick:

If we have a vector in both V and $V^\perp$, then that vector is $\vec{0}$.

Equating

$$\vec{v}_1 + \vec{w}_1 = \vec{v}_2 + \vec{w}_2,$$

rearrange to get

$$\underbrace{\vec{v}_1 - \vec{v}_2}_{\in V} = \underbrace{\vec{w}_2 - \vec{w}_1}_{\in V^\perp}.$$

But subspaces are closed under vector subtraction; therefore, the left hand side is in V whereas the right hand side is in $V^\perp$. This implies

$$\begin{aligned} \vec{v}_1 - \vec{v}_2 &\in V \\ \vec{v}_1 - \vec{v}_2 &\in V^\perp \end{aligned}$$

So by our trick:

$$\vec{v}_1 - \vec{v}_2 = \vec{0}.$$

Hence,

$$\vec{v}_1 = \vec{v}_2.$$

Likewise,

$$\begin{aligned} \vec{w}_2 - \vec{w}_1 &\in V \\ \vec{w}_2 - \vec{w}_1 &\in V^\perp \end{aligned}$$

gives us

$$\vec{w}_1 = \vec{w}_2. \qquad \qquad \square$$

This unique decomposition allows us to define a function f by

$$f(\vec{x}) = \vec{v}$$

where $\vec{v}$ is the component vector in V.

Particularly, in our Algebra II example,

$$f\begin{pmatrix} 3 \\ 4 \end{pmatrix} = \begin{bmatrix} 3 \\ 0 \end{bmatrix}$$

for

$$V = \text{span}\{\vec{e}_1\}.$$

What do we know about function f? Obviously, its image is in V:

$$f(\vec{x}) \in V$$

We also know that

$$\vec{x} - \vec{v} = \vec{w}$$

so the difference of an input and its output value under f is the orthogonal component:

$$\underbrace{\vec{x} - f(\vec{x})}_{\vec{w}} \in V^\perp$$

In fact, I claim that these two properties alone *uniquely* define the function f:

Theorem. *For any subspace $V \subseteq \mathbb{R}^n$, there is **only one** function $f : \mathbb{R}^n \to V$ that satisfies*

$$f(\vec{x}) \in V$$

and

$$\vec{x} - f(\vec{x}) \in V^\perp.$$

Proof Summary:

- Consider two such maps f_1, f_2.
- Show the difference $f_1(\vec{x}) - f_2(\vec{x})$ is in V.
- Show the difference $f_1(\vec{x}) - f_2(\vec{x})$ is in $V^\perp$.
- Conclude $f_1(\vec{x}) - f_2(\vec{x}) = \vec{0}$.

Proof: Suppose there are two functions f_1 and f_2 such that for any $\vec{x} \in \mathbb{R}^n$,

$$\begin{aligned} f_1(\vec{x}) &\in V \\ \vec{x} - f_1(\vec{x}) &\in V^\perp \end{aligned}$$

$$\begin{aligned} f_2(\vec{x}) &\in V \\ \vec{x} - f_2(\vec{x}) &\in V^\perp \end{aligned}$$

To prove two functions are equal, we need to show for any input, both functions have the same output.

Let $\vec{x} \in \mathbb{R}^n$ be an arbitrary input. If we can show

$$f_1(\vec{x}) - f_2(\vec{x}) \in V \quad \text{and} \quad f_1(\vec{x}) - f_2(\vec{x}) \in V^\perp,$$

then we know

$$f_1(\vec{x}) - f_2(\vec{x}) = \vec{0}$$

since only $\vec{0}$ lies in both V and $V^\perp$. Hence,

$$f_1(\vec{x}) = f_2(\vec{x}).$$

- $f_1(\vec{x}) - f_2(\vec{x}) \in V$

Each term is in V. Since V is closed under vector subtraction,

$$\underbrace{f_1(\vec{x})}_{\in V} - \underbrace{f_2(\vec{x})}_{\in V} \in V.$$

- $f_1(\vec{x}) - f_2(\vec{x}) \in V^{\perp}$

Add zero to rewrite the expression:

$$
\begin{aligned}
f_1(\vec{x}) - f_2(\vec{x}) &= f_1(\vec{x}) + \underbrace{(-\vec{x} + \vec{x})}_{=0} - f_2(\vec{x}) \\
&= -\left(\vec{x} - f_1(\vec{x})\right) + \left(\vec{x} - f_2(\vec{x})\right)
\end{aligned}
$$

Each summand is in $V^{\perp}$. By closure,

$$\underbrace{-\left(\vec{x} - f_1(\vec{x})\right)}_{\in V^{\perp}} + \underbrace{\left(\vec{x} - f_2(\vec{x})\right)}_{\in V^{\perp}} \in V^{\perp} \qquad \square$$

Since these two properties define a unique function, we can give this function a name:

Definition. *For any subspace $V \subseteq \mathbb{R}^n$, the **projection map***

$$P_V$$

is the unique function that satisfies

$$P_V(\vec{x}) \in V$$

and

$$\vec{x} - P_V(\vec{x}) \in V^{\perp}$$

for every $\vec{x} \in \mathbb{R}^n$.

11.6 Orthogonal Projection Properties

Like 007's car, P_V has some *sweet* features. And they will all be derived with the same trick:

> Math Mantra: To prove that two quantities are equal, it suffices to prove that
> their difference is zero.

In the universe of projection maps, we have two ways to create zero:

- To show a vector is $\vec{0}$, show that it lies in both V and $V^\perp$

- To show a dot product is 0, show that one term is in V and the other is in $V^\perp$

First we prove that P_V satisfies our favorite property:

Theorem. *For any subspace $V \subseteq \mathbb{R}^n$, $P_V(x)$ is a linear function.*

Proof Summary

- *Additive:*

  - Show $P_V(\vec{x} + \vec{y}) - P_V(\vec{x}) - P_V(\vec{y}) \in V$
  - Show $P_V(\vec{x} + \vec{y}) - P_V(\vec{x}) - P_V(\vec{y}) \in V^\perp$
  - Conclude $P_V(\vec{x} + \vec{y}) - P_V(\vec{x}) - P_V(\vec{y}) = \vec{0}$

- *Scaling*

  - Show $P_V(k\vec{x}) - kP_V(\vec{x}) \in V$
  - Show $P_V(k\vec{x}) - kP_V(\vec{x}) \in V^\perp$
  - Conclude $P_V(k\vec{x}) - kP_V(\vec{x}) = \vec{0}$

Proof:

- *Additive*

  We want to show that for any $\vec{x}, \vec{y} \in \mathbb{R}^n$,

  $$P_V(\vec{x} + \vec{y}) = P_V(\vec{x}) + P_V(\vec{y})$$

  This is equivalent to proving

  $$P_V(\vec{x} + \vec{y}) - P_V(\vec{x}) - P_V(\vec{y}) = \vec{0}$$

  Immediately

  $$\underbrace{P_V(\vec{x} + \vec{y})}_{\in V} - \underbrace{P_V(\vec{x})}_{\in V} - \underbrace{P_V(\vec{y})}_{\in V} \in V$$

  since the output of P_V is always in V, and V is closed under vector addition.

  To show this expression is also in $V^\perp$, all we need to do is add and subtract the "input" vector:

  $$P_V(\vec{x} + \vec{y}) - P_V(\vec{x}) - P_V(\vec{y}) = P_V(\vec{x} + \vec{y}) + \underbrace{\left(-(x+y) + (x+y)\right)}_{=0} - P_V(\vec{x}) - P_V(\vec{y})$$

  $$= \underbrace{-\left((\vec{x} + \vec{y}) - P_V(\vec{x} + \vec{y})\right)}_{\in V^\perp} + \underbrace{\left(x - P_V(\vec{x})\right)}_{\in V^\perp} + \underbrace{\left(\vec{y} - P_V(\vec{y})\right)}_{\in V^\perp}$$

Since $V^\perp$ is closed under vector addition,

$$P_V(\vec{x} + \vec{y}) - P_V(\vec{x}) - P_V(\vec{y}) \in V^\perp$$

Because this vector is in both V and $V^\perp$,

$$P_V(\vec{x} + \vec{y}) - P_V(\vec{x}) - P_V(\vec{y}) = \vec{0}$$

- *Scaling*
  We want to show that for any $k \in \mathbb{R}$ and $\vec{x} \in \mathbb{R}^n$,

$$P_V(k\vec{x}) = kP_V(\vec{x})$$

or equivalently,

$$P_V(k\vec{x}) - kP_V(\vec{x}) = \vec{0}.$$

By closure,

$$\underbrace{P_V(k\vec{x})}_{\in V} - k \underbrace{P_V(\vec{x})}_{\in V} \in V.$$

Again add and subtract the "input" vector:

$$P_V(k\vec{x}) - kP_V(\vec{x}) = P_V(k\vec{x}) \underbrace{-k\vec{x} + k\vec{x}}_{=\vec{0}} -kP_V(\vec{x})$$

$$= \underbrace{-\left(k\vec{x} - P_V(k\vec{x})\right)}_{\in V^\perp} + k \underbrace{\left(\vec{x} - P_V(\vec{x})\right)}_{\in V^\perp}$$

Thus,

$$P_V(k\vec{x}) - kP_V(\vec{x}) \in V^\perp.$$

Since this vector is in V and $V^\perp$, we conclude

$$P_V(k\vec{x}) - kP_V(\vec{x}) = \vec{0}. \qquad \qquad \square$$

The second cool property is that a projection map can be "swapped across" a dot product:

Theorem. *For any subspace $V \subseteq \mathbb{R}^n$ and any $\vec{x}, \vec{y} \in \mathbb{R}^n$,*

$$\vec{x} \cdot P_V(\vec{y}) = P_V(\vec{x}) \cdot \vec{y}$$

Proof Summary

- Add zero to $\vec{x} \cdot P_V(\vec{y}) - P_V(\vec{x}) \cdot \vec{y}$

- Rewrite as a sum of dot products.

- Show each dot product is between an element in V and an element in $V^\perp$.

- Conclude sum is 0.

Proof: To show

$$\vec{x} \cdot P_V(\vec{y}) - P_V(\vec{x}) \cdot \vec{y} = 0,$$

just add 0:

$$\vec{x} \cdot P_V(\vec{y}) - P_V(\vec{x}) \cdot \vec{y} = \vec{x} \cdot P_V(\vec{y}) - \underbrace{P_V(\vec{x})P_V(\vec{y}) + P_V(\vec{x})P_V(\vec{y})}_{=0} - P_V(\vec{x}) \cdot \vec{y}$$

$$= \left(x - P_V(\vec{x})\right) \cdot P_V(\vec{y}) - P_V(\vec{x}) \cdot \left(\vec{y} - P_V(\vec{y})\right)$$

But,

$$\underbrace{\left(x - P_V(\vec{x})\right)}_{\in V^{\perp}} \cdot \underbrace{P_V(\vec{y})}_{\in V} - \underbrace{P_V(\vec{x})}_{\in V} \cdot \underbrace{\left(\vec{y} - P_V(\vec{y})\right)}_{\in V^{\perp}} = 0 + 0 = 0.$$

In conclusion,

$$\vec{x} \cdot P_V(\vec{y}) - P_V(\vec{x}) \cdot \vec{y} = 0. \qquad \square$$

Lastly, we prove a "minimal distance" property. Namely, the projected vector is the element in V *closest* to the original vector.

Theorem. *For any $\vec{x} \in \mathbb{R}^n$, and $\vec{v} \in V$*

$$\|\vec{x} - P_V(\vec{x})\| \leq \|\vec{x} - \vec{v}\|$$

Moreover,

$$\|\vec{x} - P_V(\vec{x})\| = \|\vec{x} - \vec{v}\| \iff \vec{v} = P_V(\vec{x})$$

Proof Summary

- Consider $\|\vec{x} - \vec{v}\|^2$.

- Add $P_V(\vec{x}) - P_V(\vec{x})$ inside the norm and expand.

- Cancel dot products between elements in V and $V^{\perp}$.

- Remaining equality implies all results.

Proof: First, we derive a useful inequality by considering the square

$$\|\vec{x} - \vec{v}\|^2.$$

First, include $P_V(\vec{x})$ terms by adding zero:

$$\|\vec{x} \underbrace{- P_V(\vec{x}) + P_V(\vec{x})}_{\vec{0}} - \vec{v}\|^2.$$

Expanding as a dot product, we get

$$\|\left(\vec{x} - P_V(\vec{x})\right) + \left(P_V(\vec{x}) - \vec{v}\right)\|^2 = \left(\underbrace{\left(\vec{x} - P_V(\vec{x})\right)}_{\vec{a}} + \underbrace{\left(P_V(\vec{x}) - \vec{v}\right)}_{\vec{b}}\right) \cdot \left(\underbrace{\left(\vec{x} - P_V(\vec{x})\right)}_{\vec{a}} + \underbrace{\left(P_V(\vec{x}) - \vec{v}\right)}_{\vec{b}}\right)$$

$$= \underbrace{\|\vec{x} - P_V(\vec{x})\|^2}_{\|\vec{a}\|^2} + \underbrace{2\left(\vec{x} - P_V(\vec{x})\right) \cdot \left(P_V(\vec{x}) - \vec{v}\right)}_{2\vec{a} \cdot \vec{b}} + \underbrace{\|P_V(\vec{x}) - \vec{v}\|^2}_{\|\vec{b}\|^2}$$

Since the inner term is zero

$$2 \underbrace{\left(\vec{x} - P_V(\vec{x})\right)}_{\in V^\perp} \cdot \left(\overbrace{P_V(\vec{x})}^{\in V} - \overbrace{\vec{v}}^{\in V} \right) = 0,$$

we have

$$\|\vec{x} - \vec{v}\|^2 = \|\vec{x} - P_V(\vec{x})\|^2 + \|P_V(\vec{x}) - \vec{v}\|^2. \qquad (\star)$$

- $\|\vec{x} - P_V(\vec{x})\| \leq \|\vec{x} - \vec{v}\|$

  The right side of $(\star)$ is a sum of non-negative terms; therefore, the total sum must bound any of its parts. In particular,

$$\|\vec{x} - \vec{v}\|^2 \geq \|\vec{x} - P_V(\vec{v})\|^2.$$

  Since square roots preserve inequalities,

$$\|\vec{x} - \vec{v}\| \geq \|\vec{x} - P_V(\vec{v})\|.$$

- $\|\vec{x} - P_V(\vec{x})\| = \|\vec{x} - \vec{v}\| \iff \vec{v} = P_V(\vec{x})$

  Suppose

$$\|\vec{x} - \vec{v}\| = \|\vec{x} - P_V(\vec{v})\|$$

  Then our equality $(\star)$ becomes

$$\underbrace{\|\vec{x} - P_V(\vec{v})\|^2}_{\|\vec{x}-\vec{v}\|^2} = \|\vec{x} - P_V(\vec{x})\|^2 + \|P_V(\vec{x}) - \vec{v}\|^2.$$

  Therefore,

$$\|P_V(\vec{x}) - \vec{v}\|^2 = 0$$

  which only happens when

$$P_V(\vec{x}) = \vec{v}.$$

  Conversely, if $P_V(\vec{x}) = \vec{v}$,

$$\|\vec{x} - P_V(\vec{v})\| = \|\vec{x} - \underbrace{\vec{v}}_{P_V(\vec{v})}\|. \qquad \square$$

New Notation

Symbol	Reading	Example	Example Translation
$V^\perp$	The orthogonal complement of V or V "perp."	$V^\perp = \{\vec{0}\}$	The orthogonal complement of V is $\{\vec{0}\}$.
$P_V(\vec{x})$	The projection of $\vec{x}$ onto V.	$P_V(\vec{x} + \vec{y}) = P_V(\vec{x}) + P_V(\vec{y})$	The projection of $\vec{x} + \vec{y}$ onto V is the sum of the projection of $\vec{x}$ onto V and the projection of $\vec{y}$ onto V.

Lecture 12

A Game of Cat and Gauss

Sometimes the truth isn't good enough, sometimes people deserve more.
Sometimes people deserve to have their faith rewarded...

-Batman, *The Dark Knight*

Goals: Using Gaussian Elimination, we construct explicit formulas for the bases of the null space and the column space. Not only is this useful in practice, but this construction also gives us an alternate proof to the Rank-Nullity Theorem. Finally, we end this unit by explaining how to solve inhomogeneous systems of equations.

12.1 A Little "Constructive" Criticism

By now, we have cited the basis theorem a gazillion times. We used it to assert the existence of bases for the null space and the column space, to derive the Rank-Nullity Theorem, and to prove that the dimensions of the column space and the row space are equal. That's great. But, in case you haven't noticed:

We never told you how to explicitly calculate these bases.

It's not enough to simply assert that the bases exist. In the real world, we *need* to be able to explicitly find the bases for the null space and the column space. How else are our Kindles, iPads, and microwaves going to work?

We *deserve* more. We deserve to have our faith in proofs rewarded. Therefore, we are going to give a rigorous method to *construct* the bases. And as an added bonus, we will get an alternate proof of the Rank-Nullity Theorem. So everyone- mathematicians, engineers, and even those crazy mathematical constructivists- will be happy.

12.2   Gaussian Elimination

Recall, from **Lecture 4**, that we can transform a system of equations

$$
\begin{array}{ccccccccc}
a_{11}x_1 & + & a_{12}x_2 & + & a_{13}x_3 & + & \dots & + & a_{1n}x_n & = & 0 \\
a_{21}x_1 & + & a_{22}x_2 & + & a_{23}x_3 & + & \dots & + & a_{2n}x_n & = & 0 \\
& \vdots & & \vdots & & \vdots & & \vdots & & \vdots \\
a_{m1}x_1 & + & a_{m2}x_2 & + & a_{m3}x_3 & + & \dots & + & a_{mn}x_n & = & 0
\end{array}
$$

into one of two possible forms $Usingmatrices, wecanrepresentthisreductioninamorecondensedform. Then$

reduced to one of two matrices:

$$
\begin{bmatrix}
0 & a'_{12} & a'_{13} & \dots & a'_{1n} \\
0 & a'_{22} & a'_{23} & \dots & a'_{2n} \\
\vdots & \vdots & \vdots & \ddots & \vdots \\
0 & a'_{m2} & a'_{m3} & \dots & a'_{mn}
\end{bmatrix}
\qquad
\begin{bmatrix}
1 & a'_{12} & a'_{13} & \dots & a'_{1n} \\
0 & a'_{22} & a'_{23} & \dots & a'_{2n} \\
\vdots & \vdots & \vdots & \ddots & \vdots \\
0 & a'_{m2} & a'_{m3} & \dots & a'_{mn}
\end{bmatrix}
$$

For ease, let's call this maneuver $(C1)$ for *first column reduction*.

We will use $(C1)$ to define **Gaussian Elimination** on matrices. Even though our algorithm may look unintuitive, just remember:

> *The reductions are the same operations used in solving **a homogeneous system of equations.** Matrices are just a **neat** shorthand to represent these operations.*

Moreover, Gaussian Elimination is a *systematic* procedure. And through this matrix shorthand, we will discover cool *patterns*.

Gaussian Elimination is composed of **two phases**.

- Phase I repeatedly applies $C1$ and records *pivot column indices*. At the end of this phase, we say that our matrix A is in *row echelon form*.

- Phase II uses the pivot column indices acquired in Phase I to completely "clean" the system. We say the final matrix is in *reduced row echelon form*.

We define Phase I inductively:

<u>GAUSSIAN ELIMINATION: PHASE I</u>

Consider the $m \times n$ matrix

$$\begin{bmatrix} a_{11} & a_{12} & a_{13} & \ldots & a_{1n} \\ a_{21} & a_{22} & a_{23} & \ldots & a_{2n} \\ \vdots & \vdots & \vdots & \ddots & \vdots \\ a_{m1} & a_{m2} & a_{m3} & \ldots & a_{mn} \end{bmatrix}$$

- **Step 1**

  Perform $(C1)$ on this matrix to get one of two cases:

  - CASE 1

    The matrix reduces to

    $$\begin{bmatrix} 0 & a'_{12} & a'_{13} & \ldots & a'_{1n} \\ 0 & a'_{22} & a'_{23} & \ldots & a'_{2n} \\ \vdots & \vdots & \vdots & \ddots & \vdots \\ 0 & a'_{m2} & a'_{m3} & \ldots & a'_{mn} \end{bmatrix}$$

    Then define S_2 to be the sub-matrix that **ignores the first column**:

    $$\begin{bmatrix} 0 & a'_{12} & a'_{13} & \ldots & a'_{1n} \\ 0 & a'_{22} & a'_{23} & \ldots & a'_{2n} \\ \vdots & \vdots & \vdots & \ddots & \vdots \\ 0 & a'_{m2} & a'_{m3} & \ldots & a'_{mn} \end{bmatrix}$$

  - CASE 2

    The matrix reduces to

    $$\begin{bmatrix} 1 & a'_{12} & a'_{13} & \ldots & a'_{1n} \\ 0 & a'_{22} & a'_{23} & \ldots & a'_{2n} \\ \vdots & \vdots & \vdots & \ddots & \vdots \\ 0 & a'_{m2} & a'_{m3} & \ldots & a'_{mn} \end{bmatrix}$$

    Record, as the first *pivot column index*:

    $$P_1 = 1.$$

    Define S_2 to be the sub-matrix that **ignores the first row and the first column**:

    $$\begin{bmatrix} 1 & a'_{12} & a'_{13} & \ldots & a'_{1n} \\ 0 & a'_{22} & a'_{23} & \ldots & a'_{2n} \\ \vdots & \vdots & \vdots & \ddots & \vdots \\ 0 & a'_{m2} & a'_{m3} & \ldots & a'_{mn} \end{bmatrix}$$

- **Step** $\ell \geq 2$

  Let

  $$
  \begin{bmatrix}
  b_{11} & b_{12} & b_{13} & \dots & b_{1n} \\
  b_{21} & b_{22} & b_{23} & \dots & b_{2n} \\
  \vdots & \vdots & \vdots & \ddots & \vdots \\
  b_{m1} & b_{m2} & b_{m3} & \dots & b_{mn}
  \end{bmatrix}
  $$

  be the full transformed matrix acquired after performing **Step** $\ell - 1$

  Focus on the sub-matrix S_ℓ:

  $$
  \begin{bmatrix}
  \dots & \dots & \vdots & \vdots & \vdots & \vdots \\
  \dots & \dots & \vdots & \vdots & \vdots & \vdots \\
  \dots & \dots & b_{i\ell} & b_{i(\ell+1)} & \dots & b_{in} \\
  \dots & \dots & b_{(i+1)\ell} & b_{(i+1)(\ell+1)} & \dots & b_{(i+1)n} \\
  \dots & \dots & \vdots & \vdots & \ddots & \vdots \\
  \dots & \dots & b_{m\ell} & b_{m(\ell+1)} & \dots & b_{mn}
  \end{bmatrix}
  $$

  Leaving the columns and equations of the larger matrix **unchanged**, perform $(C1)$ on sub-matrix S_ℓ. Again, we have two cases:

  - CASE 1

    Sub-matrix S_ℓ reduces to:

    $$
    \begin{bmatrix}
    \dots & \dots & \vdots & \vdots & \vdots & \vdots \\
    \dots & \dots & \vdots & \vdots & \vdots & \vdots \\
    \dots & \dots & 0 & b'_{i(\ell+1)} & \dots & b'_{in} \\
    \dots & \dots & 0 & b'_{(i+1)(\ell+1)} & \dots & b'_{(i+1)n} \\
    \dots & \dots & \vdots & \vdots & \ddots & \vdots \\
    \dots & \dots & 0 & b'_{m(\ell+1)} & \dots & b'_{mn}
    \end{bmatrix}
    $$

    Then define $S_{\ell+1}$ to be the sub-matrix that ignores the **first column of** S_ℓ:

    $$
    \begin{bmatrix}
    \dots & \dots & \vdots & \vdots & \vdots & \vdots \\
    \dots & \dots & \vdots & \vdots & \vdots & \vdots \\
    \dots & \dots & 0 & b'_{i(\ell+1)} & \dots & b'_{in} \\
    \dots & \dots & 0 & b'_{(i+1)(\ell+1)} & \dots & b'_{(i+1)n} \\
    \dots & \dots & \vdots & \vdots & \ddots & \vdots \\
    \dots & \dots & 0 & b'_{m(\ell+1)} & \dots & b'_{mn}
    \end{bmatrix}
    $$

- CASE 2

Sub-matrix S_ℓ reduces to

$$\begin{bmatrix} \cdots & \cdots & \vdots & \vdots & & \vdots & & \vdots \\ \cdots & \cdots & \vdots & \vdots & & \vdots & & \vdots \\ \cdots & \cdots & 1 & b'_{i(\ell+1)} & \cdots & b'_{in} \\ \cdots & \cdots & 0 & b'_{(i+1)(\ell+1)} & \cdots & b'_{(i+1)n} \\ \cdots & \cdots & \vdots & \vdots & \ddots & \vdots \\ \cdots & \cdots & 0 & b'_{m(\ell+1)} & \cdots & b'_{mn} \end{bmatrix}$$

Given that steps $1, 2, \ldots$ produced pivot column indices

$$P_1, P_2, \ldots P_k,$$

we record the next pivot column index to be the *current column number* ℓ.

$$P_{k+1} = \ell$$

Define $S_{\ell+1}$ to be the sub-matrix that ignores the **first column and first row of S_ℓ**:

$$\begin{bmatrix} \cdots & \cdots & \vdots & \vdots & & \vdots & & \vdots \\ \cdots & \cdots & \vdots & \vdots & & \vdots & & \vdots \\ \cdots & \cdots & 1 & b'_{i(\ell+1)} & \cdots & b'_{in} \\ \cdots & \cdots & 0 & b'_{(i+1)(\ell+1)} & \cdots & b'_{(i+1)n} \\ \cdots & \cdots & \vdots & \vdots & \ddots & \vdots \\ \cdots & \cdots & 0 & b'_{m(\ell+1)} & \cdots & b'_{mn} \end{bmatrix}$$

- Phase I is **finished** once sub-system S_ℓ is empty.

Upon completing Phase I, the resulting matrix is in *row echelon form matrix*. Moreover, we have a list of pivot columns indices:

$$P_1, P_2, \ldots P_Q.$$

The columns with these indices are called *pivot columns*:

$$\begin{bmatrix} & | & & | & & | & \\ \cdots & \vec{\alpha}_{P_1} & \cdots & \vec{\alpha}_{P_2} & \cdots & \vec{\alpha}_{P_Q} & \cdots \\ & | & & | & & | & \end{bmatrix}$$

Out of convention, we label (in order) the indices of the remaining $n - Q$ columns

$$N_1, N_2, \ldots, N_{n-Q}$$

and call the corresponding columns *non-pivot columns*:

$$
\begin{bmatrix}
| & & | & & | \\
\vec{\alpha}_{N_1} & \cdots & \vec{\alpha}_{N_2} & \cdots & \vec{\alpha}_{N_{n-Q}} \\
| & & | & & |
\end{bmatrix}
$$

All together, the columns of our matrix are labelled[1]:

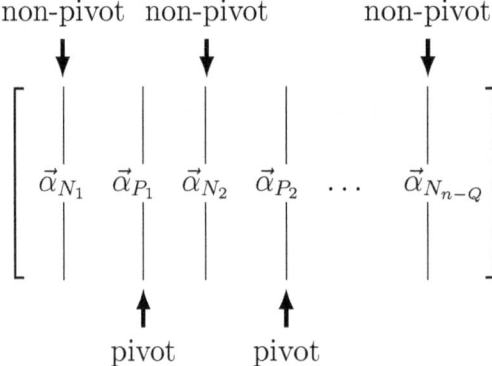

Moreover, each *pivot column* $\vec{a}_{P_i}$ is of the form

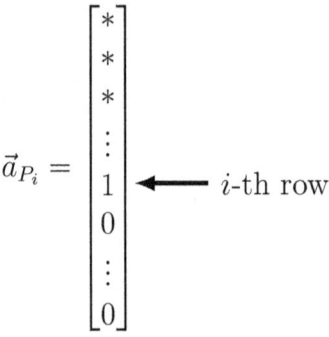

where the $*$'s are numbers that need not be zero. This can be rigorously proven with induction. Although we omit the proof, this fact is simply a consequence **Case 2**. To see this, let's write out the first few steps of Phase I:

We start with the full matrix as the sub-matrix:

$$
\begin{bmatrix}
\cdots & \cdots & \cdots & \cdots & \cdots & \cdots \\
\cdots & \cdots & \cdots & \cdots & \cdots & \cdots \\
\cdots & \cdots & \cdots & \cdots & \cdots & \cdots \\
\cdots & \cdots & \cdots & \cdots & \cdots & \cdots \\
\cdots & \cdots & \cdots & \cdots & \cdots & \cdots
\end{bmatrix}
$$

---

[1]Be careful when interpreting the following schematic! The non-pivot and pivot columns need not be in alternating order. This diagram merely emphasizes that *every* column is labelled.

Columns are successively removed from this sub-matrix (Case 1)

$$\begin{bmatrix} \cdots & \cdots & \cdots & \cdots & \cdots & \cdots \\ \cdots & \cdots & \cdots & \cdots & \cdots & \cdots \\ \cdots & \cdots & \cdots & \cdots & \cdots & \cdots \\ \cdots & \cdots & \cdots & \cdots & \cdots & \cdots \\ \cdots & \cdots & \cdots & \cdots & \cdots & \cdots \end{bmatrix} \longrightarrow \begin{bmatrix} 0 & \cdots & \cdots & \cdots & \cdots & \cdots \\ 0 & \cdots & \cdots & \cdots & \cdots & \cdots \\ 0 & \cdots & \cdots & \cdots & \cdots & \cdots \\ \vdots & \cdots & \cdots & \cdots & \cdots & \cdots \\ 0 & \cdots & \cdots & \cdots & \cdots & \cdots \end{bmatrix} \longrightarrow \begin{bmatrix} 0 & 0 & \cdots & \cdots & \cdots & \cdots \\ 0 & 0 & \cdots & \cdots & \cdots & \cdots \\ 0 & 0 & \cdots & \cdots & \cdots & \cdots \\ \vdots & \vdots & \cdots & \cdots & \cdots & \cdots \\ 0 & 0 & \cdots & \cdots & \cdots & \cdots \end{bmatrix}$$

until we enter Case 2. Here, we select the first pivot column as the first column of a sub-matrix that has **no rows removed**.

$$\begin{matrix} & & P_1 \\ & & \downarrow \\ \begin{bmatrix} 0 & \cdots & 1 & \cdots & \cdots & \cdots \\ 0 & \cdots & 0 & \cdots & \cdots & \cdots \\ 0 & \cdots & 0 & \cdots & \cdots & \cdots \\ \vdots & \cdots & \vdots & \cdots & \cdots & \cdots \\ 0 & \cdots & 0 & \cdots & \cdots & \cdots \end{bmatrix} \end{matrix}$$

Then the first row (and column) is removed to form a new sub-matrix:

$$\begin{bmatrix} 0 & \cdots & 1 & \cdots & \cdots & \cdots \\ 0 & \cdots & 0 & \cdots & \cdots & \cdots \\ 0 & \cdots & 0 & \cdots & \cdots & \cdots \\ \vdots & \cdots & \vdots & \cdots & \cdots & \cdots \\ 0 & \cdots & 0 & \cdots & \cdots & \cdots \end{bmatrix}$$

Continuing, columns are successively removed from this sub-matrix (Case 1)

$$\begin{bmatrix} 0 & \cdots & 1 & \cdots & \cdots & \cdots \\ 0 & \cdots & 0 & \cdots & \cdots & \cdots \\ 0 & \cdots & 0 & \cdots & \cdots & \cdots \\ \vdots & \cdots & \vdots & \cdots & \cdots & \cdots \\ 0 & \cdots & 0 & \cdots & \cdots & \cdots \end{bmatrix} \longrightarrow \begin{bmatrix} 0 & \cdots & 1 & \cdots & \cdots & \cdots \\ 0 & \cdots & 0 & 0 & \cdots & \cdots \\ 0 & \cdots & 0 & 0 & \cdots & \cdots \\ \vdots & \cdots & \vdots & \vdots & \cdots & \cdots \\ 0 & \cdots & 0 & 0 & \cdots & \cdots \end{bmatrix} \longrightarrow \begin{bmatrix} 0 & \cdots & 1 & \cdots & \cdots & \cdots \\ 0 & \cdots & 0 & 0 & 0 & \cdots \\ 0 & \cdots & 0 & 0 & 0 & \cdots \\ \vdots & \cdots & \vdots & \vdots & \vdots & \cdots \\ 0 & \cdots & 0 & 0 & 0 & \cdots \end{bmatrix}$$

until the second pivot column is selected (Case 2):

$$\begin{matrix} & P_1 & & P_2 \\ & \downarrow & & \downarrow \\ \begin{bmatrix} \cdots & 1 & \cdots & * & \cdots & \cdots \\ \cdots & 0 & \cdots & 1 & \cdots & \cdots \\ \cdots & 0 & \cdots & 0 & \cdots & \cdots \\ \cdots & \vdots & \cdots & \vdots & \cdots & \cdots \\ \cdots & 0 & \cdots & 0 & \cdots & \cdots \end{bmatrix} \end{matrix}$$

Then we go through the same process until the third pivot column is selected:

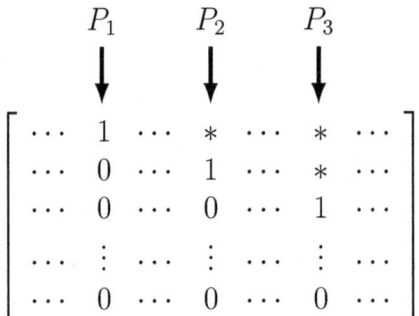

If you do not fully understand Phase I, go back and read it again. Once you fully understand it, you can continue to Phase II.

The second phase simply "kills" the values above the 1 in each pivot column:

GAUSSIAN ELIMINATION: PHASE II

After performing Phase I, we are left with a matrix in *row echelon form* where the pivot columns are exactly columns $P_1, P_2, \ldots, P_Q$:

$$
\begin{bmatrix}
 & | & & | & & | & \\
\cdots & \vec{\alpha}_{P_1} & \cdots & \vec{\alpha}_{P_2} & \cdots & \vec{\alpha}_{P_Q} & \cdots \\
 & | & & | & & | &
\end{bmatrix}
$$

For each pivot column $\vec{\alpha}_{P_i}$,

$$
\begin{array}{c}
P_i \\
\downarrow
\end{array}
$$

$$
\begin{bmatrix}
\cdots & * & \cdots \\
\cdots & * & \cdots \\
\cdots & * & \cdots \\
\cdots & \vdots & \cdots \\
\cdots & 1 & \cdots \\
\cdots & 0 & \cdots \\
\cdots & \vdots & \cdots \\
\cdots & 0 & \cdots
\end{bmatrix} i
$$

from each row above the i-th, subtract the right multiple of the i-th row so that the entry in column P_i becomes 0:

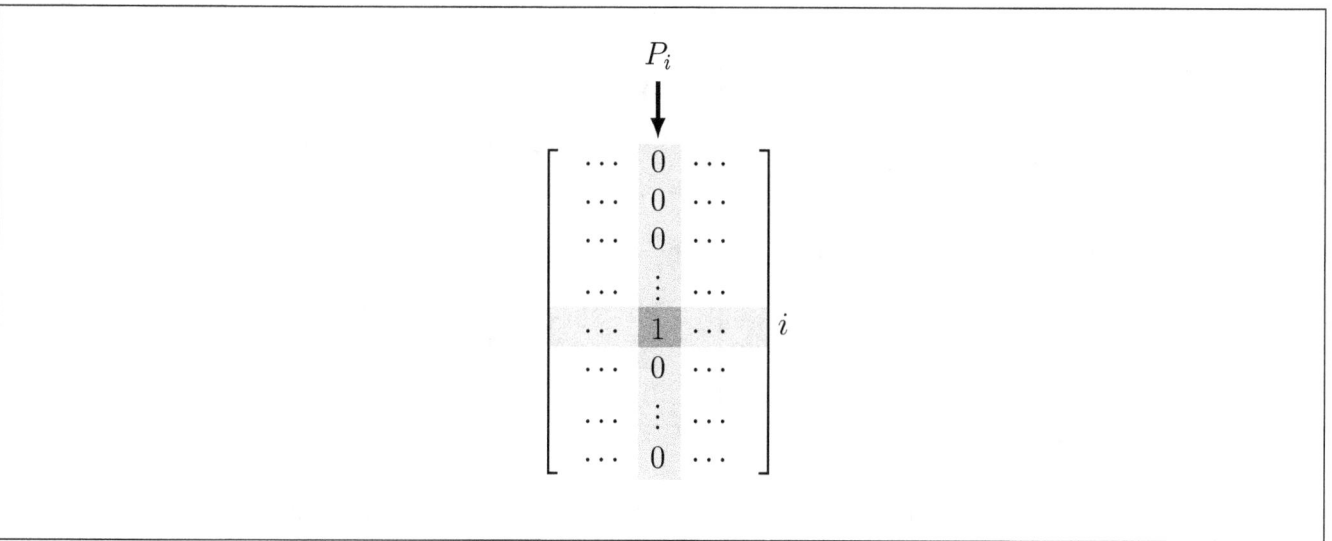

Formally,

Definition. *After performing Gaussian Elimination on a matrix A, the resulting matrix is denoted*

$$\mathrm{rref}(A)$$

*and is called the **reduced row echelon form of A**.*

Example. *Perform Gaussian Elimination on the matrix*

$$A = \begin{bmatrix} 2 & 4 & 2 & 6 \\ 2 & 4 & 2 & 6 \\ 2 & 4 & 4 & 10 \\ 2 & 4 & 0 & 2 \end{bmatrix}$$

- PHASE ONE

  – **Step I**

  Apply $(C1)$ to get

  $$\begin{bmatrix} 1 & 2 & 1 & 3 \\ 0 & 0 & 0 & 0 \\ 0 & 0 & 2 & 4 \\ 0 & 0 & -2 & -4 \end{bmatrix}$$

  Since we are in Case 2, set

  $$P_1 = 1$$

  and define the sub-matrix S_2 by ignoring the first row and first column:

$$\begin{bmatrix} 1 & 2 & 1 & 3 \\ 0 & 0 & 0 & 0 \\ 0 & 0 & 2 & 4 \\ 0 & 0 & -2 & -4 \end{bmatrix}$$

- **Step 2**

  Apply $(C1)$ on the sub-matrix S_2 to get

$$\begin{bmatrix} 1 & 2 & 1 & 3 \\ 0 & 0 & 0 & 0 \\ 0 & 0 & 2 & 4 \\ 0 & 0 & -2 & -4 \end{bmatrix}$$

  Since we are in Case 1, define sub-matrix S_3 by ignoring the first column of S_2:

$$\begin{bmatrix} 1 & 2 & 1 & 3 \\ 0 & 0 & 0 & 0 \\ 0 & 0 & 2 & 4 \\ 0 & 0 & -2 & -4 \end{bmatrix}$$

- **Step 3**

  Apply $(C1)$ on the sub-matrix S_3 to get

$$\begin{bmatrix} 1 & 2 & 1 & 3 \\ 0 & 0 & 1 & 2 \\ 0 & 0 & 0 & 0 \\ 0 & 0 & 0 & 0 \end{bmatrix}$$

  Since we are in Case 2, set

$$P_2 = 3$$

  and define sub-matrix S_4 by ignoring the first column and first row of S_3:

$$\begin{bmatrix} 1 & 2 & 1 & 3 \\ 0 & 0 & 1 & 2 \\ 0 & 0 & 0 & 0 \\ 0 & 0 & 0 & 0 \end{bmatrix}$$

- **Step 4**

  Apply $(C1)$ on the first column of sub-matrix S_4 to get

$$\begin{bmatrix} 1 & 2 & 1 & 3 \\ 0 & 0 & 1 & 2 \\ 0 & 0 & 0 & 0 \\ 0 & 0 & 0 & 0 \end{bmatrix}$$

Since we are in Case 1, define sub-matrix S_5 by ignoring the first column of S_4. Since S_5 is empty, we are done with Phase I.

- PHASE II
  From Phase I, we have pivot column indices

$$
\begin{aligned}
P_1 &= 1 \\
P_2 &= 3
\end{aligned}
$$

and row echelon form

$$
\begin{bmatrix}
1 & 2 & 1 & 3 \\
0 & 0 & 1 & 2 \\
0 & 0 & 0 & 0 \\
0 & 0 & 0 & 0
\end{bmatrix}
$$

Kill the entry above the 1 in pivot column P_2 by subtracting the second row from the first:

$$
\text{rref}(A) = \begin{bmatrix}
1 & 2 & 0 & 1 \\
0 & 0 & 1 & 2 \\
0 & 0 & 0 & 0 \\
0 & 0 & 0 & 0
\end{bmatrix}
$$

One question you should be asking yourself is,

What is the point of Gaussian Elimination?

Notice that this procedure *reflects* the solution preserving-operations used to solve a homogeneous systems of equations. This means that

The solution set of

$$
A\vec{x} = 0
$$

*is **the same** as the solution set of*

$$
\text{rref}(A)\vec{x} = 0
$$

i.e.

$$
N(A) = N\big(\text{rref}(A)\big)
$$

By studying the structure of a rref(A), we will derive an explicit formula for a basis of $N(A)$. Generally,

> Math Mantra: By reducing an object to some basic canonical form, we can exploit
> its structural properties.

But,

What is so special about reduced row echelon form?

Matrices in reduced row echelon form have an **incredible** structural property:[1]

The i-th pivot column of rref(A) **is the i-th standard basis vector!**

---

[1]This can be proven rigorously with induction. If you have doubts, complete it during Thanksgiving break.

Precisely,

$$\vec{b}_{P_i} = \vec{e}_i$$

where $\vec{b}_k$ is the k-th column of rref(A).

In our example, the first and second pivot columns are $\vec{e}_1$ and $\vec{e}_2$ respectively:

$$\begin{bmatrix} 1 & 2 & 0 & 3 \\ 0 & 0 & 1 & 2 \\ 0 & 0 & 0 & 0 \\ 0 & 0 & 0 & 0 \end{bmatrix}$$

As for non-examples, the following matrices are not in reduced row echelon form:

$$\begin{bmatrix} 1 & 0 & 0 & 0 & 0 \\ 0 & 0 & 0 & 0 & 0 \\ 0 & 0 & 1 & 0 & 0 \\ 0 & 0 & 0 & 1 & 0 \\ 0 & 0 & 0 & 0 & 0 \end{bmatrix}, \quad \begin{bmatrix} 1 & 0 & 0 & 0 & 0 \\ 0 & 1 & 0 & 0 & 0 \\ 0 & 0 & 0 & 0 & 1 \\ 0 & 0 & 0 & 1 & 0 \\ 0 & 0 & 0 & 0 & 0 \end{bmatrix}$$

The left matrix fails because it is missing $\vec{e}_2$ whereas the right matrix fails because the standard basis are out of order.

12.3   An Enlightening Example

The explicit formula for the[1] null space basis looks very **intimidating**. But, to quote Professor Simon,

It's just book-keeping!

In order to make the formula and its derivation *precise*, we need heavy notation. Particularly, we need to use double subscripts to distinguish a *pivot column* from a *non-pivot column*. But the proof is really just **simple algebra in disguise!**

In order to *motivate* the proof, we are going to do a specific example.[2] Generally,

> Math Mantra: We can find inspiration for theorems in examples.

Then, we will prove the theorem in the special case when the pivots are actually the first Q columns. This is because of another mantra,

---

[1]We will often say "the" to describe the basis that we will construct. Of course, the null space has many bases. But this method finds one particular "special" basis.

[2]For the midterm, you shouldn't even memorize the explicit null space basis formula. You should just understand how to find a null space basis in any particular problem by following the steps in our example.

> Math Mantra: If you cannot solve the harder problem, make a simplifying
> assumption and try to solve the easier version.[1]

Finally, when you are ready to attack the notation, we will prove the general statement.

Example. *Find a basis for the null space of the matrix*

$$A = \begin{bmatrix} 1 & 6 & 1 & 0 & 15 \\ 0 & 0 & 1 & 1 & 17 \\ 0 & 0 & 0 & 2 & 18 \\ 2 & 12 & 0 & 0 & 14 \end{bmatrix}$$

Use Phase I of Gaussian Elimination to reduce A to row echelon form and find the pivot indices:

$$\begin{aligned} P_1 &= 1 \\ P_2 &= 3 \\ P_3 &= 4 \end{aligned}$$

It follows that the non-pivot indices are:

$$\begin{aligned} N_1 &= 2 \\ N_2 &= 5 \end{aligned}$$

Then use Phase II to reduce A to reduced row echelon form

$$\text{rref}(A) = \begin{bmatrix} 1 & 6 & 0 & 0 & 7 \\ 0 & 0 & 1 & 0 & 8 \\ 0 & 0 & 0 & 1 & 9 \\ 0 & 0 & 0 & 0 & 0 \end{bmatrix}$$

Expand

$$\text{rref}(A)\vec{x} = \vec{0}$$

as a system of equations

$$\begin{aligned} x_1 + 6x_2 + 7x_5 &= 0 \\ x_3 + 8x_5 &= 0 \\ x_4 + 9x_5 &= 0 \end{aligned}$$

Isolate:

$$\begin{aligned} x_1 &= -6x_2 - 7x_5 \\ x_3 &= -8x_5 \\ x_4 &= -9x_5 \end{aligned}$$

---

[1] *Tons* of special cases of Fermat's Last Theorem cases were verified long before Andrew Wiles came along.

This means for $\vec{x} \in N(A)$,

$$\vec{x} = \begin{bmatrix} x_1 \\ x_2 \\ x_3 \\ x_4 \\ x_5 \end{bmatrix} = \begin{bmatrix} -6x_2 - 7x_5 \\ x_2 \\ -8x_5 \\ -9x_5 \\ x_5 \end{bmatrix} = \begin{bmatrix} -6x_2 \\ x_2 \\ 0 \\ 0 \\ 0 \end{bmatrix} + \begin{bmatrix} -7x_5 \\ 0 \\ -8x_5 \\ -9x_5 \\ x_5 \end{bmatrix}$$

$$= x_2 \begin{bmatrix} -6 \\ 1 \\ 0 \\ 0 \\ 0 \end{bmatrix} + x_5 \begin{bmatrix} -7 \\ 0 \\ -8 \\ -9 \\ 1 \end{bmatrix}$$

Since x_2, x_5 can take on any values, the null space is

$$N(A) = \text{span} \left\{ \begin{bmatrix} -6 \\ 1 \\ 0 \\ 0 \\ 0 \end{bmatrix}, \begin{bmatrix} -7 \\ 0 \\ -8 \\ -9 \\ 1 \end{bmatrix} \right\}$$

In more condensed notation,

$$N(A) = \text{span} \{ \vec{e}_2 - 6\vec{e}_1, \quad \vec{e}_5 - 7\vec{e}_1 - 8\vec{e}_3 - 9\vec{e}_4 \}$$

Stare at the indices of the standard basis vectors:

$$\vec{e}_2 - 6\vec{e}_1, \quad \vec{e}_5 - 7\vec{e}_1 - 8\vec{e}_3 - 9\vec{e}_4$$

and notice that these are the non-pivot and pivot indices:

$$\vec{e}_{N_1} - 6\vec{e}_{P_1}, \quad \vec{e}_{N_2} - 7\vec{e}_{P_1} - 8\vec{e}_{P_2} - 9\vec{e}_{P_3}$$

Moreover, the scaling coefficients are just the entries of the reduced row echelon form:

$$\vec{e}_{N_1} - 6\vec{e}_{P_{\boxed{1}}} \qquad \vec{e}_{N_2} - 7\vec{e}_{P_{\boxed{1}}} - 8\vec{e}_{P_{\boxed{2}}} - 9\vec{e}_{P_{\boxed{3}}}$$

Generally, to construct each vector in the null space basis, we

- Take a standard basis vector with *non-pivot index*

$$\vec{e}_{N_j}$$

and look at the corresponding non-pivot column $\vec{b}_{N_j}$ in rref(A):

$$\begin{bmatrix} b_{1N_j} \\ b_{2N_j} \\ \vdots \\ b_{mN_j} \end{bmatrix}$$

- For **each** of the pivot columns

$$P_1, P_2, \ldots, P_Q$$

subtract the standard basis with index P_i scaled by the i-th entry in column $\vec{b}_{N_j}$:

$$\vec{e}_{N_j} - b_{1N_j}\vec{e}_{P_1} - b_{2N_j}\vec{e}_{P_2} - \ldots - b_{QN_j}\vec{e}_{P_Q}$$

This is an awesome result, but proving it is *simple*: it is just **basic algebra in disguise**. In fact, the proof is merely a *formalization* of our enlightening example.

12.4   An Easier Theorem

Before we derive the full-blown explicit null space basis formula, let's do an easier case. This way, you can understand the general idea of the proof before juggling double subscript notation.

Theorem. *Let A be an $m \times n$ matrix and $B = $ rref(A). Moreover, assume*

$$1, 2, \ldots, Q$$

are the indices of the pivot columns and

$$Q+1, Q+2, \ldots, n$$

are the indices of the non-pivot columns. Then the basis for the null space of A is

$$\vec{e}_{Q+1} \quad - \quad \sum_{i=1}^{Q} b_{i(Q+1)}\vec{e}_i,$$

$$\vec{e}_{Q+2} \quad - \quad \sum_{i=1}^{Q} b_{i(Q+2)}\vec{e}_i,$$

$$\vdots$$

$$\vec{e}_n \quad - \quad \sum_{i=1}^{Q} b_{in}\vec{e}_i,$$

Proof Summary:

- *Spanning*

  - View $B\vec{x} = \vec{0}$ as a system of equations.

  - Look at the equations corresponding to pivot column numbers

  - Since our pivot columns are standard basis vectors, all but one pivot column variable remains in each equation.

  - Isolate that variable.

  - Rewrite the solution vector $\vec{x}$ in terms of non-pivot column variables.

  - Separate the vector into basis components.

- *Linear Independence*

  - Each $\vec{e}_j - \sum\limits_{i=1}^{Q} b_{in}\vec{e}_i$ uniquely has a 1 in its j-th component, where $j \geq Q+1$.

Proof:

- *Spanning*

  Let $\vec{x}$ be a solution to

  $$B\vec{x} = \vec{0}$$

  Expand the system of equations and look at the equations corresponding to the pivot components (the first Q equations):

  $$\begin{array}{ccccccccc}
  b_{11}x_1 & + & b_{12}x_2 & + & \cdots & b_{1Q}x_Q & \cdots & + & b_{1n}x_n & = & 0 \\
  b_{21}x_1 & + & b_{22}x_2 & + & \cdots & b_{2Q}x_Q & \cdots & + & b_{2n}x_n & = & 0 \\
  \vdots & & \vdots & & & \vdots & & & \vdots & & \vdots \\
  b_{Q1}x_1 & + & b_{Q2}x_2 & + & \cdots & b_{QQ}x_Q & \cdots & + & b_{Qn}x_n & = & 0
  \end{array} \qquad (\star)$$

  Recall that the pivot columns are **standard basis vectors**. In particular, the first Q columns of B are the first Q standard basis vectors:

  $$\begin{array}{cc}
  \overbrace{}^{\text{Pivot}} & \overbrace{\phantom{b_{1(Q+1)} \; \cdots \; b_{1n}}}^{\text{Non-pivot}} \\
  \end{array}$$

  $$\begin{bmatrix}
  1 & 0 & 0 & \cdots & b_{1(Q+1)} & \cdots & b_{1n} \\
  0 & 1 & 0 & \cdots & b_{2(Q+1)} & \cdots & b_{2n} \\
  0 & 0 & 1 & \cdots & b_{3(Q+1)} & \cdots & b_{3n} \\
  \vdots & \vdots & \vdots & \cdots & \vdots & & \\
  0 & 0 & 0 & \cdots & b_{mn} & \cdots & b_{mn}
  \end{bmatrix}$$

  This means, for column $j = 1, 2, \ldots, Q$, we have

  $$b_{ij} = \begin{cases} 0 & i \neq j \\ 1 & i = j \end{cases}$$

Therefore, our system $(\star)$ becomes

$$
\begin{aligned}
x_1 + b_{1(Q+1)}x_{(Q+1)} + b_{1(Q+2)}x_{(Q+2)} + \ldots + b_{1n}x_n &= 0 \\
x_2 + b_{2(Q+1)}x_{(Q+1)} + b_{2(Q+2)}x_{(Q+2)} + \ldots + b_{2n}x_n &= 0 \\
\vdots \qquad\qquad \vdots \qquad\qquad\qquad \vdots \qquad \vdots \qquad\qquad \vdots \\
x_Q + b_{Q(Q+1)}x_{Q(Q+1)} + b_{Q(Q+2)}x_{(Q+2)} + \ldots + b_{Qn}x_n &= 0
\end{aligned}
$$

Isolating $x_1, x_2, \ldots x_Q$, we get the constraints

$$
\begin{aligned}
x_1 &= -\big(b_{1(Q+1)}x_{(Q+1)} + b_{1(Q+2)}x_{(Q+2)} + \ldots + b_{1n}x_n\big) \\
x_2 &= -\big(b_{2(Q+1)}x_{(Q+1)} + b_{2(Q+2)}x_{(Q+2)} + \ldots + b_{2n}x_n\big) \\
\vdots &\qquad\qquad \vdots \qquad\qquad\qquad \vdots \qquad \vdots \\
x_Q &= -\big(b_{Q(Q+1)}x_{Q(Q+1)} + b_{Q(Q+2)}x_{(Q+2)} + \ldots + b_{Qn}x_n\big)
\end{aligned}
$$

Substituting into $\vec{x}$

$$
\begin{bmatrix} x_1 \\ x_2 \\ \vdots \\ x_Q \\ x_{Q+1} \\ \vdots \\ x_N \end{bmatrix}
=
\begin{bmatrix}
-\big(b_{1(Q+1)}x_{(Q+1)} + b_{1(Q+2)}x_{(Q+2)} + \ldots + b_{1n}x_n\big) \\
-\big(b_{2(Q+1)}x_{(Q+1)} + b_{2(Q+2)}x_{(Q+2)} + \ldots + b_{2n}x_n\big) \\
\vdots \qquad\qquad \vdots \qquad \vdots \\
-\big(b_{Q(Q+1)}x_{Q(Q+1)} + b_{Q(Q+2)}x_{(Q+2)} + \ldots + b_{Qn}x_n\big) \\
x_{Q+1} \\
\vdots \\
x_N
\end{bmatrix}
$$

$$
=
\begin{bmatrix} -b_{1(Q+1)}x_{Q+1} \\ -b_{2(Q+1)}x_{Q+1} \\ \vdots \\ -b_{Q(Q+1)}x_{Q+1} \\ x_{Q+1} \\ 0 \\ \vdots \\ 0 \end{bmatrix}
+
\begin{bmatrix} -b_{1(Q+2)}x_{Q+2} \\ -b_{2(Q+2)}x_{Q+2} \\ \vdots \\ -b_{Q(Q+2)}x_{Q+2} \\ 0 \\ x_{Q+2} \\ \vdots \\ 0 \end{bmatrix}
+ \ldots +
\begin{bmatrix} -b_{1n}x_n \\ -b_{2n}x_n \\ \vdots \\ -b_{Qn}x_n \\ 0 \\ 0 \\ \vdots \\ x_n \end{bmatrix}
$$

$$
=
x_{Q+1}\begin{bmatrix} -b_{1(Q+1)} \\ -b_{2(Q+1)} \\ \vdots \\ -b_{Q(Q+1)} \\ 1 \\ 0 \\ \vdots \\ 0 \end{bmatrix}
+
x_{Q+2}\begin{bmatrix} -b_{1(Q+2)} \\ -b_{2(Q+2)} \\ \vdots \\ -b_{Q(Q+2)} \\ 0 \\ 1 \\ \vdots \\ 0 \end{bmatrix}
+ \ldots +
x_n\begin{bmatrix} -b_{1n} \\ -b_{2n} \\ \vdots \\ -b_{Qn} \\ 0 \\ 0 \\ \vdots \\ 1 \end{bmatrix}
$$

Since $x_{Q+1}, x_{Q+2}, \ldots, x_n$ can be any value,

$$\vec{e}_{Q+1} \ - \ \sum_{i=1}^{Q} b_{i(Q+1)}\vec{e}_i,$$

$$\vec{e}_{Q+2} \ - \ \sum_{i=1}^{Q} b_{i(Q+2)}\vec{e}_i,$$

$$\vdots$$

$$\vec{e}_n \ - \ \sum_{i=1}^{Q} b_{in}\vec{e}_i,$$

span the null space.

- *Linear Independence*
  When expanded, each vector

$$\vec{e}_j - \sum_{i=1}^{Q} b_{ij}\vec{e}_i$$

uniquely has 1 at its j-th component:

$$\begin{bmatrix} -b_{1(Q+1)} \\ -b_{2(Q+1)} \\ \vdots \\ -b_{Q(Q+1)} \\ 1 \\ 0 \\ \vdots \\ 0 \end{bmatrix}, \begin{bmatrix} -b_{1(Q+2)} \\ -b_{2(Q+2)} \\ \vdots \\ -b_{Q(Q+2)} \\ 0 \\ 1 \\ \vdots \\ 0 \end{bmatrix}, \dots, \begin{bmatrix} -b_{1n} \\ -b_{2n} \\ \vdots \\ -b_{Qn} \\ 0 \\ 0 \\ \vdots \\ 1 \end{bmatrix}$$

$\square$

12.5   Null Space Basis

If you understand the previous proof, then the following proof is just optional. You won't be tested on it and it's just simple algebra stated in *precise notation*. In fact, the only difference between this proof and the previous one is that we consider a *sequence* of pivot indices. This forces us to introduce double subscripts.

Theorem. *Let* $B = \mathrm{rref}(A)$ *and let*

$$P_1, P_2, \ldots, P_Q$$

be the indices of the pivot columns and[1]

$$N_1, N_2, \ldots, N_K$$

be indices of the non-pivot columns. Then the basis for the null space of A is

$$\vec{e}_{N_1} \quad - \quad \sum_{i=1}^{Q} b_{iN_1} \vec{e}_{P_i},$$

$$\vec{e}_{N_2} \quad - \quad \sum_{i=1}^{Q} b_{iN_2} \vec{e}_{P_i},$$

$$\vdots$$

$$\vec{e}_{N_K} \quad - \quad \sum_{i=1}^{Q} b_{iN_k} \vec{e}_{P_i},$$

Proof Summary:

- *Spanning*

    - View $B\vec{x} = \vec{0}$ as a system of equations.
    - Look at the equations corresponding to pivot column numbers
    - Since our pivot columns are standard basis vectors, all but one pivot column variable remains in each equation.
    - Isolate that variable.
    - Rewrite the solution vector $\vec{x}$ in terms of non-pivot column variables.
    - Separate the vector into basis components.

- *Linear Independence*

    - Each $\vec{e}_{N_j} - \sum_{i=1}^{Q} b_{iN_j} \vec{e}_{P_i}$ uniquely has a 1 in its N_j-th component.

Proof:

- *Spanning*
  Let $\vec{x}$ to a solution to

$$B\vec{x} = \vec{0}$$

---

[1] K is, of course, $n - Q$. We only leave it as K for simplification.

and look at the equations of the expanded system. The i-th equation tells us

$$\sum_{j=1}^{n} b_{ij} x_j = 0.$$

Split this sum into two. Each sum will have standard basis vectors corresponding to non-pivot and pivot columns, respectively:

$$\underbrace{\sum_{r=1}^{Q} b_{iP_r} x_{P_r}}_{\text{Pivots}} + \underbrace{\sum_{r=1}^{K} b_{iN_r} x_{N_r}}_{\text{Non-Pivots}} = 0 \qquad (\star)$$

By definition,

$$b_{iP_r}$$

is the i-th entry of the r-th **pivot column**. Moreover, we know the r-th pivot column is the r-th standard basis vector

$$\vec{b}_{P_r} = \vec{e}_r$$

Therefore,

$$b_{iP_r} = \begin{cases} 1 & \text{if } i = r \\ 0 & \text{otherwise} \end{cases}$$

Thus,

$$\sum_{r=1}^{Q} b_{iP_r} x_{P_j} = \overbrace{b_{iP_1} x_{P_1} + b_{iP_2} x_{P_2} + \dots}^{0} + \overbrace{b_{iP_i} x_{P_i}}^{x_{P_i}} + \overbrace{\dots + b_{iP_{Q-1}} x_{P_{Q-1}} + b_{iP_Q} x_{P_Q}}^{0}$$

and the i-th equation $(\star)$ collapses into

$$x_{P_i} + \underbrace{\sum_{r=1}^{K} b_{iN_r} x_{N_r}}_{\text{Non-Pivots}} = 0$$

i.e

$$x_{P_i} = -\left(\sum_{r=1}^{K} b_{iN_r} x_{N_r} \right).$$

Split $\vec{x}$ into its separate standard basis components

$$\vec{x} = \sum_{i=1}^{n} x_i \vec{e}_i$$

Again, separate this sum into

$$\vec{x} = \underbrace{\sum_{i=1}^{K} x_{N_i} \vec{e}_{N_i}}_{\text{Non-pivots}} + \underbrace{\sum_{i=1}^{Q} x_{P_i} \vec{e}_{P_i}}_{\text{Pivots}}.$$

Substituting ($\star$) into each pivot variable, we rewrite $\vec{x}$ as

$$\vec{x} = \sum_{i=1}^{K} x_{N_i} \vec{e}_{N_i} + \sum_{i=1}^{Q} \underbrace{-\left(\sum_{r=1}^{K} b_{iN_r} x_{N_r} \right)}_{x_{P_i}} \vec{e}_{P_i}$$

i.e

$$\vec{x} = \sum_{i=1}^{K} x_{N_i} \vec{e}_{N_i} - \sum_{i=1}^{Q} \sum_{r=1}^{K} b_{iN_r} x_{N_r} \vec{e}_{P_i}$$

Since double sums commute, this is

$$\vec{x} = \sum_{i=1}^{K} x_{N_r} \vec{e}_{N_r} - \sum_{r=1}^{K} \sum_{i=1}^{Q} b_{iN_i} x_{N_i} \vec{e}_{P_i}$$

Then, pull out the constant x_{N_r} from the inner sum

$$\vec{x} = \sum_{i=1}^{K} x_{N_i} \vec{e}_{N_i} - \sum_{r=1}^{K} x_{N_r} \left(\sum_{i=1}^{Q} b_{iN_r} \vec{e}_{P_i} \right)$$

and change a dummy variable from i to r:

$$\vec{x} = \sum_{r=1}^{K} x_{N_r} \vec{e}_{N_r} - \sum_{r=1}^{K} x_{N_r} \left(\sum_{i=1}^{Q} b_{iN_r} \vec{e}_{P_i} \right)$$

Collapse using *distributive law*:

$$\vec{x} = \sum_{r=1}^{K} x_{N_r} \left(\vec{e}_{N_r} - \sum_{i=1}^{Q} b_{iN_r} \vec{e}_{P_i} \right).$$

But x_{N_r} can take on any value. Therefore, the null space is the span of

$$\vec{e}_{N_1} - \sum_{i=1}^{Q} b_{iN_1} \vec{e}_{P_i},$$

$$\vec{e}_{N_2} - \sum_{i=1}^{Q} b_{iN_2} \vec{e}_{P_i},$$

$$\vdots$$

$$\vec{e}_{N_K} - \sum_{i=1}^{Q} b_{iN_K} \vec{e}_{P_i},$$

- *Linear Independence*
  If we look at the N_j-th component

$$\left[\vec{e}_{N_j} - \sum_{i=1}^{Q} b_{iN_j} \vec{e}_{P_i} \right]_{N_j}$$

we see that its value is 1:

$$\overbrace{\left[\vec{e}_{N_j}\right]_{N_j}}^{1} - \overbrace{\left[b_{1N_j}\vec{e}_{P_1}\right]_{N_j}}^{0} - \overbrace{\left[b_{2N_j}\vec{e}_{P_2}\right]_{N_j}}^{0} - \ldots - \overbrace{\left[b_{QN_j}\vec{e}_{P_Q}\right]_{N_j}}^{0} = 1$$

Moreover, for $k \neq j$, the N_j-th component

$$\left[\vec{e}_{N_k} - \sum_{i=1}^{Q} b_{iN_k}\vec{e}_{P_i}\right]_{N_j}$$

have value 0:

$$\overbrace{\left[\vec{e}_{N_k}\right]_{N_j}}^{0} - \overbrace{\left[b_{1N_k}\vec{e}_{P_1}\right]_{N_j}}^{0} - \overbrace{\left[b_{2N_k}\vec{e}_{P_2}\right]_{N_j}}^{0} - \ldots - \overbrace{\left[b_{QN_k}\vec{e}_{P_Q}\right]_{N_j}}^{0} = 0$$

Thus, each vector

$$\vec{e}_{N_j} - \sum_{i=1}^{Q} b_{iN_j}\vec{e}_{P_i}$$

is the only vector with 1 in its N_j-th component. □

12.6   Column Space Basis

Because of the *massive* amount of book-keeping we did in constructing the null space basis, we are rewarded for our troubles. We can actually use the null space basis to *derive* an explicit formula for the column space basis. In fact, we even get a pithy description:

The basis of the column space of A is the columns of A corresponding to the pivot indices.

Note that we are referring to the columns of the **original** matrix, not the reduced row echelon form.[1]

Theorem. *Let A be an $m \times n$ matrix with columns $\vec{\alpha}_i$:*

$$\begin{bmatrix} | & | & & | \\ \vec{\alpha}_1 & \vec{\alpha}_2 & \ldots & \vec{\alpha}_n \\ | & | & & | \end{bmatrix}$$

Let $B = \mathrm{rref}(A)$ with pivot indices

$$P_1, P_2, \ldots, P_Q$$

and non-pivot indices

$$N_1, N_2, \ldots, N_K.$$

Then,

$$\vec{\alpha}_{P_1}, \vec{\alpha}_{P_2}, \ldots, \vec{\alpha}_{P_Q}$$

is a basis for A.

---

[1]That would be nuttier than squirrel poo: it would mean that every vector space is a span of standard basis vectors!

Proof Summary:

- *Spanning*:

  - $\subseteq$

    Obvious

  - $\supseteq$

    * Suffices to show an arbitrary non-pivot column is in the span of the pivot columns.
    * Multiply A by this non-pivot columns' corresponding null space basis vector.
    * Rewrite product as the non-pivot column in terms of pivot columns.

- *Linear Independence*:

  - Suppose not, then linear dependence gives us a null space vector $\vec{c}$.
  - Write $\vec{c}$ as a combination of the explicit null space basis vectors.
  - $\vec{c}$ is zero at some non-pivot index while the combination is non-zero at the same location. Contradiction.

Proof:

- *Spanning*

  Obviously,

  $$\text{span}\left\{\vec{\alpha}_{P_1}, \vec{\alpha}_{P_2}, \ldots \vec{\alpha}_{P_Q}\right\} \subseteq \text{span}\left\{\vec{\alpha}_1, \vec{\alpha}_2, \ldots, \vec{\alpha}_n\right\}$$

  Therefore, we need only show $\supseteq$. Moreover, it suffices to show that any of the other columns of A is contained in

  $$\text{span}\left\{\vec{\alpha}_{P_1}, \vec{\alpha}_{P_2}, \ldots \vec{\alpha}_{P_Q}\right\}$$

  Consider a column that is not a pivot column. By definition, it is enumerated by some non-pivot index:

  $$\vec{\alpha}_{N_i}$$

  Then, take the null space basis vector

  $$\vec{e}_{N_i} - \sum_{i=1}^{Q} b_{iN_i} \vec{e}_{P_i}$$

  and multiply it by A:

  $$A\left(\vec{e}_{N_i} - \sum_{i=1}^{Q} b_{iN_i} \vec{e}_{P_i}\right) = 0$$

  This yields

  $$A\left(\vec{e}_{N_i}\right) = A\left(\sum_{i=1}^{Q} b_{iN_1} \vec{e}_{P_i}\right)$$

  But multiplying A by some standard basis vector $\vec{e}_j$ simply picks out the j-th column of A. Therefore,

  $$\vec{\alpha}_{N_i} = \sum_{i=1}^{Q} b_{iN_1} \vec{\alpha}_{P_i}$$

i.e,

$$\vec{\alpha}_{N_i} \in \text{span}\left\{\vec{\alpha}_{P_1}, \vec{\alpha}_{P_2}, \ldots \vec{\alpha}_{P_Q}\right\}$$

- *Linear Independence*
  Suppose they are not. Then we can find a non-trivial solution

$$c_{P_1}\vec{\alpha}_{P_1} + c_{P_2}\vec{\alpha}_{P_2} + \ldots + c_{P_Q}\vec{\alpha}_{P_Q} = \vec{0}$$

But this gives us a non trivial vector in the null space. Precisely, we have a matrix multiplication with a vector $\vec{c}$ where

$$\vec{c} = \sum_{i=1}^{Q} c_{P_i}\vec{e}_{P_i}.$$

Visually,

$$\begin{bmatrix} & | & & | & & | & \\ \cdots & \vec{\alpha}_{P_1} & \cdots & \vec{\alpha}_{P_2} & \cdots & \vec{\alpha}_{P_Q} & \cdots \\ & | & & | & & | & \end{bmatrix} \begin{bmatrix} \vdots \\ c_{P_1} \\ \vdots \\ c_{P_2} \\ \vdots \\ c_{P_Q} \\ \vdots \end{bmatrix} = \vec{0}$$

Using our explicit formula for the null space basis, we know that

$$\vec{c} = s_{N_1}\left(\vec{e}_{N_1} - \sum_{i=1}^{Q} b_{iN_1}\vec{e}_{P_i}\right) + \ldots + s_{N_r}\left(\vec{e}_{N_r} - \sum_{i=1}^{Q} b_{iN_r}\vec{e}_{P_i}\right) + \ldots + s_{N_K}\left(\vec{e}_{N_K} - \sum_{i=1}^{Q} b_{iN_K}\vec{e}_{P_i}\right)$$

where at least one s_{N_r} is non-zero. In particular, notice that the (RHS) is non-zero at its N_r component:

$$\underbrace{\left[s_{N_1}\left(\vec{e}_{N_1} - \sum_{i=1}^{Q} b_{iN_1}\vec{e}_{P_i}\right)\right]_{N_r}}_{=0} + \ldots + \underbrace{\left[s_{N_r}\left(\vec{e}_{N_r} - \sum_{i=1}^{Q} b_{iN_r}\vec{e}_{P_i}\right)\right]_{N_r}}_{\neq 0} + \ldots + \underbrace{\left[s_{N_K}\left(\vec{e}_{N_K} - \sum_{i=1}^{Q} b_{iN_K}\vec{e}_{P_i}\right)\right]}_{=0}$$

However, by definition, $\vec{c}$ is 0 at its N_r component! So we have a contradiction. $\square$

Notice that our *explicit* basis formulas give us an alternate proof of the Rank-Nullity theorem. We explicitly constructed a null space basis that has as many vectors as there are pivot columns, hence:

$$\dim N(A) = \text{\# of pivot columns}.$$

And we explicitly constructed a column space basis that has as many vectors as there are non-pivot columns, hence:

$$\dim C(A) = \text{\# of non-pivot columns}.$$

Of course, every column is either a pivot or non-pivot:

$$\text{\# of non-pivot columns} + \text{\# of pivot columns} = \text{\# of columns}.$$

Therefore,

$$\underbrace{\dim N(A)}_{\text{\# of pivot columns}} + \underbrace{\dim C(A)}_{\text{\# of non-pivot columns}} = \underbrace{n}_{\text{\# of columns}}$$

12.7 Inhomogeneous Equations

We end this lecture with some pretty easy proofs. These proofs are of practical importance and after three weeks of intense math, you earned a break.

In the real world, we would like to solve inhomogeneous systems of equations of the form

$$A\vec{x} = \vec{b}.$$

Particularly, we need to ask,

1. How do we know there *exists* a solution?

2. How do we find *one* solution?

3. How do we find *all* solutions?

The first question is easy. The product

$$A\vec{x}$$

is a *linear combination* of columns. In fact, a solution exists if and only if $\vec{b}$ is in the column space:

$$A\vec{x} = \vec{b} \text{ has a solution if and only if } \vec{b} \in C(A).$$

Practically, we check this the same way that we did in high school. Consider the inhomogeneous system:

$$\begin{array}{ccccc} x_1 & + & x_2 & = & 1 \\ x_1 & + & x_2 & = & 5 \end{array}$$

After applying Gaussian Elimination, we have the equivalent system

$$\begin{array}{ccccc} x_1 & + & x_2 & = & 1 \\ 0 & + & 0 & = & 4 \end{array}$$

Here, we can instantly see that there is no solution (otherwise we would have 0=4).

This idea can be extended to arbitrary matrices. First, we can easily prove that the solutions to

$$A\vec{x} = \vec{b}$$

are the same as the solutions to

$$A'\vec{x} = \vec{b'}$$

where $A', \vec{b'}$ are the sub-matrices acquired from performing Gaussian Elimination on an *augmented* matrix:[1]

$$\text{rref}\left[A\middle|\vec{b}\right] = \left[A'\middle|\vec{b'}\right]$$

Moreover, we can prove that a solution exists if and only if for every row of zeros in A', the corresponding component of $\vec{b'}$ is also zero:

---

[1]I'm not going into too much detail since you've already learned this procedure in high school. Everything is precisely the same except your *perspective* has changed.

$$\begin{bmatrix} \vdots & \vdots & \vdots & \vdots & \vdots \\ \vdots & \vdots & \vdots & \vdots & \vdots \\ 0 & 0 & \dots & 0 & 0 \\ \vdots & \vdots & \vdots & \vdots & \vdots \end{bmatrix}$$

We can also use Gaussian Elimination to answer the second question. To construct a *particular* solution, perform Gaussian Elimination on the augmented matrix. Then, substitute arbitrary values into variables with non-pivot indices. In fact, it is easiest to set

$$\begin{aligned} x_{N_1} &= 0 \\ x_{N_2} &= 0 \\ &\vdots \\ x_{N_K} &= 0 \end{aligned}$$

We can then directly solve for each of the pivot variables.

Example. *Find a particular solution $\vec{x}_0$ to*

$$A\vec{x} = \vec{b}$$

where

$$A = \begin{bmatrix} 1 & 2 & 0 & 1 \\ 0 & 0 & 1 & 2 \\ 0 & 0 & 0 & 0 \\ 0 & 0 & 0 & 0 \end{bmatrix}$$

and

$$\vec{b} = \begin{bmatrix} 4 \\ 3 \\ 0 \\ 0 \end{bmatrix}.$$

We want to solve the system

$$\begin{aligned} x_1 \;+\; 2x_2 \;+\; x_4 &= 4 \\ x_3 \;+\; 2x_4 &= 3 \end{aligned}$$

Since A is already in reduced row echelon form,

$$\begin{aligned} N_1 &= 2 \\ N_2 &= 4 \end{aligned}$$

Therefore, set

$$\begin{aligned} x_2 &= 0 \\ x_4 &= 0 \end{aligned}$$

to get

$$x_1 = 4$$
$$x_3 = 3$$

Thus,

$$\vec{x}_0 = \begin{bmatrix} 4 \\ 0 \\ 3 \\ 0 \end{bmatrix}$$

is a particular solution.

Once you have a particular solution to an inhomogeneous system, you immediately know *all solutions*. It is simply the set of all null space vectors *shifted*[1] by the particular solution:

Theorem. *Let $\vec{x}_0$ be a particular solution to*

$$A\vec{x} = \vec{b}. \tag{$\star$}$$

Then, $\vec{x}$ is a solution of ($\star$) if and only if $\vec{x} \in N(A) + \vec{x}_0$ where

$$N(A) + \vec{x}_0 = \{\vec{n} + \vec{x}_0 \mid \vec{n} \in N(A)\}$$

Proof: Let $\vec{x}_0$ be a particular solution:

$$A\vec{x}_0 = \vec{b}.$$

- $\Rightarrow$

  For any $\vec{x}$ such that

  $$A\vec{x} = \vec{b}$$

  we have

  $$A\vec{x} - \vec{b} = \vec{0}.$$

  Substituting, we get

  $$A\vec{x} - \underbrace{A\vec{x}_0}_{\vec{b}} = \vec{0}$$

  Therefore,

  $$A(\vec{x} - \vec{x}_0) = \vec{0}$$

  so

  $$(\vec{x} - \vec{x}_0) \in N(A).$$

  Since

  $$\vec{x} = \underbrace{(\vec{x} - \vec{x}_0)}_{\in N(A)} + \vec{x}_0$$

  we conclude

  $$\vec{x} \in N(A) + \vec{x}_0$$

---

[1] We typically call the sum of a vector and subspace an *affine subspace*.

- $\Leftarrow$

Let $\vec{x} \in N(A) + \vec{x}_0$. Then

$$\vec{x} = \vec{n} + \vec{x}_0$$

for some $\vec{n} \in N(A)$. Multiplying,

$$A\vec{x} = A\left(\vec{n} + \vec{x}_0\right) = \underbrace{A\vec{n}}_{\vec{0}} + \underbrace{A\vec{x}_0}_{\vec{b}} = \vec{b}. \qquad \square$$

New Notation

Symbol	Reading	Example	Example Translation
$\mathrm{rref}(A)$	The reduced row echelon form of the matrix A	$N(A) = N\left(\mathrm{rref}(A)\right)$	The null space of A equals the null space of the reduced row echelon form of A.

Midterm I: The Linear Algebra Menace

If you only know definitions, theorem statements, and computation, then...
YOU SHALL NOT PASS

- Gand𝐱

Making a Choice and Mastering the Material

If you don't like doing proofs, then you gotta get out. **Now.** Transfer to Math 51. There's no shame in doing so. However, if you still feel in your bones that you *need* to be a mathematician, then stay. But unless you want to be trampled by all those IMO, SUMaC, PROMYS, CMS math prodigies, you need to master the material. But how do you even *know* whether you've obtained such mastery? There are two age-old tests:

1. **Can you re-derive the proofs from scratch?**
   It is not enough to be able to read it from a book. To quote Professor Devlin,

   > *It's like learning to ride a bike. Someone can ride up and down in front of you for hours, telling you how they do it. But you won't learn to ride from watching them and having them explain it to you. You have to keep trying for yourself and FAILING until it eventually clicks.*

2. **Can you explain it out loud?**
   Find someone, a dorm member, an upperclassman, or even a fellow Math 51H compatriot, and reteach the proofs. To quote Hilbert,

   > *A mathematical theory is not to be considered complete until you have made it so clear that you can explain it to the first man whom you meet on the street.*

   Or, if you are too shy, go to a vacant classroom with a chalkboard and start talking to yourself. Or even better, just don a ninja costume and start teaching on YouTube.

Ask yourself the following questions to see if you have mastered all the topics.

Week 1

1. Do you understand how to do basis proofs?

   - "Arbitrary" Proofs
   - Proof by Contradiction
   - Proof by Cases
   - If and Only If
   - Induction

2. Do you know the definition of the distance and the angle between two (non-zero) vectors?

3. Do you understand the geometry of vectors? Can you construct line segments and lines using sets of vectors?

4. Do you feel comfortable working with inequalities: squaring and square rooting, establishing upper bounds?

5. Do you know all the dot product properties and **how to prove them**?

6. Do you know the fundamental relation between norm and dot product? Can you prove the law of cosines, the parallelogram law, etc.?

7. **Can you state and prove the Cauchy-Schwarz inequality?**

8. **Can you state and prove the Cauchy-Schwarz equality?**

9. **Can you apply Cauchy-Schwarz to prove other inequalities?**

10. Do you know the definition of a linear function?

11. Do you know the definition of a subspace? **Can you verify that a given set is a subspace?**

12. **Can you prove all the linear independence/dependence properties and equivalences?**

13. **Can you state and prove the Under-determined Systems Lemma?**

14. **Can you state and prove the Linear Dependence Lemma?**

Week 2

1. Do you know how to prove something is unique?

2. Can you prove statements using the Field Axiom?

3. Do you understand the *abstract meaning* of a *rational* number?

4. Do you understand the *abstract meaning* of a *real* number?

5. **Can you state the completeness axiom?** Can you use the Completeness Axiom to prove that some number exists? Can you prove that number satisfies some property?

6. Can you prove two sets are equal?

7. **Can you state the Basis Theorem and prove it?**

8. **Can you state the Basis Extension Theorem and prove it?**

9. **Can you prove that dimension is unique?**

10. Can you prove dimension properties?

11. Can you prove two matrices are equal?

12. Do you know that $A\vec{x}$ is a linear combination of the column vectors of A?

13. Do you understand the norm of a matrix and can you prove the "Cauchy Schwarz-like bound?"

14. Can you solve for the matrix A that represents a given linear transformation?

15. Do you understand null space, column space, and row space?

16. **Can you state and prove the Rank-Nullity Theorem?**

17. Can you prove rank properties?

Week 3

1. **Do you know the definition of a limit? Can you prove that a sequence converges to a specific number?**

2. **Can you prove the limit properties?**

3. **Can you state and prove the Monotone Convergence Property?**

4. **Can you state and prove the Sandwich Theorem?**

5. **Can you state and prove the Bolzano-Weierstrass Theorem?**

6. **Can you state and prove all the orthogonal complement properties?**

7. **Can you state and prove all the projection map properties?**

8. **Can you prove that the dimension of the column space and the dimension of the row space are equal?**

9. **Can you calculate bases for the null space and the column space?**

10. **Can you solve for all the solutions of an inhomogeneous system of equations? Can you define an affine space?**

Final Advice

That's the content. Here's some final pointers that will make a *huge difference*:

1. **DO THE PRACTICE TEST.**
   The questions on the real test will be similar to the questions on the practice test.

2. **MAKE SURE YOUR DEFINITIONS ARE EXACT.**
   Professor Simon is pretty relentless about this: if the question asks you to give a definition, your statement must be flawless.

3. **GET ACCUSTOMED TO WORKING AT 7PM.**
   The test is at 7PM. I know. It sucks. So do practice questions (or even the practice test) at this time.

Alright, I think that's everything. This is going to be the first *real* math test you take: barely any calculation, 1 or 2 definition questions, and the rest **proofs**. So, to quote the Hunger Games,

Good luck, and may the odds be ever in your favor.